THE NATURAL MAN

THE NATURAL MAN

ED McCLANAHAN

Farrar, Straus and Giroux
NEW YORK
1983
229p

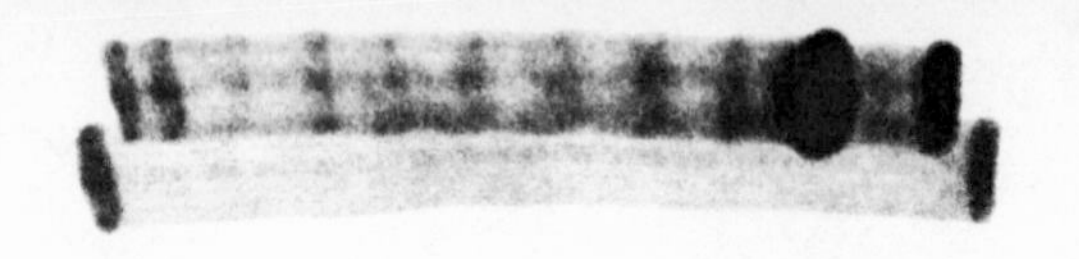

Sections of this book appeared, in different form, in *Esquire*, *Place*, and *Adena*.

The characters and events herein are fictitious and any similarity to persons living or dead is purely coincidental.

Printed in the United States of America

Published simultaneously in Canada by
McGraw-Hill Ryerson Ltd., Toronto

Designed by Tere LoPrete

FIRST PRINTING, 1983

Library of Congress Cataloging in Publication Data

McClanahan, Ed.
The natural man.

I. Title.
PS3563.C3397N3 1983 813'.54 82-21056

This book is for

WILLIE GORDON RYAN

Long may he wave

You are of the class, mammalia; order, primates; genus, homo; species, Kentucky.

—COOPER, *The Prairie*

THE NATURAL MAN

PROLOGUE

In the days of his youth, in the summertime, when school didn't interfere, Harry Eastep liked to spend a certain portion of his afternoons hanging around Marvin Conklin's drugstore, to catch the arrival of the Cincinnati–Lexington Greyhound. It wasn't that he was meeting anyone or going anywhere—scarcely anyone who needed meeting ever came to Needmore, and Harry scarcely ever went away. But the drugstore was the drop station for Needmore's daily supply of the Cincinnati *Morning Enquirer*, so the bus, which delivered the paper to the provinces, always stopped there. Nor was Harry more than most boys interested in the *Enquirer*'s already half-stale accounts of the affairs of whatever great world it was that lay beyond the narrow confines of Burdock County. The fact is, it was the bus driver himself that Harry came to see.

There were always two or three loafers sitting around in Conklin's, drinking fountain Cokes, when the bus roared up and stopped in the middle of the street between the drugstore and the Burdock County courthouse across the way. As if to inform the local yokels that he didn't intend to stay any longer than he damn well had to, the driver never bothered to cut off his engine; the bus sat there throbbing and muttering in flatulent impatience while the driver alighted through the fold-back door and came striding smartly across the street, a smallish banty rooster of a fellow, slim-hipped and natty in his tailored uniform, his cap cocked low and rakish on his brow, his nifty little Mandrake the Magician mustache already twitching in pleasurable anticipation, the thick log of newspapers balanced one-handed on his epauletted shoulder. Then, flinging open Marvin Conklin's screen door, he'd cry, in perfect rubbernecker tour-guide lookit-the-funny-natives-folks singsong, complete with an Ohio accent, "State of Kentucky, County of Burdock, City of Needmore, population 6⅞ when they're all at home! Where prosperity is *a-a-always* just around the corner! *Heads up, hayseeds!*" With that, he'd heave the roll of papers end over end into the store, where it landed at the feet of the foremost loafer with a dreadful thud that shook the Coke glasses and patent-medicine bottles on the shelves; and the bus driver's scornful, mustachioed grin would hang there in the doorway till the screen door clapped shut after him and dispelled it. Moments later, as the Greyhound thundered off in a fulsome cloud of blue exhaust, the dumb-struck loafers would turn at last to one another, grousing bitterly about that high-handed little wisenheimer's latest effrontery. "Why, that smart-aleck snot!" they'd remind each other. "That cocky little s'rimp! That wise guy, that twerp, that . . ."

But to young Harry Eastep, lurking about the magazine rack in the drugstore's dimly lighted recesses, the bus driver was the hero, not the villain, of this brief yet curiously satisfying little matinee; he had, well, class, Harry guessed you'd call it; he was *suavay*. Harry didn't feel the least bit suavay himself, of course. As a matter of fact, he sometimes imagined himself to be the very object of the driver's scorn, the most egregiously obvious hayseed of all the 357 hayseed souls in Needmore. Harry, in the wisdom of his fifteen years, had come to suspect that there was a good deal more to the world than had so far met his eye, and to discover within himself a powerful itch to sally forth and See What It Was

All About, to blow this one-horse jerkwater burg (as he imagined the bus driver might put it) and go where the Action was.

By that summer near the end of the 1940's, Harry Eastep and Needmore, Kentucky, had had the dubious pleasure of each other's company for six long years. Population 67⁄8, and Harry had been among that wretched number since his parents took him out of the fifth grade of the Wilbur Wright Elementary School in Dayton, Ohio, and trundled him down to Needmore with them to live with his aged, newly widowed, grim-visaged grandmother, "Miss" Lute Biddle, for the duration—theirs or hers. Miss Lute proved nothing if not durable, and six years later there they sat.

Needmore had changed not a hair in those six years, so far as Harry could determine; Harry, meanwhile, had changed most marvelously, from a plump, bespectacled, modestly precocious nine-year-old (an only child, with working parents and nothing better to do, he'd started school early, done his homework, and skipped the second grade) to a slouching, shambling, gangling tangle of ganglia, an almost-senior at Burdock County High, not yet sixteen, still bespectacled and modestly precocious, though not half so smart as he thought he was. Nearly six feet tall, he wore his height uneasily, like a new recruit in an ill-fitting uniform. He still looked up—literally and figuratively—to people who no longer needed looking up to, a habit which did little to improve his slouch or enhance his bearing, and indeed had caused more than one observer to liken his expression to that of a shit-eating dog. But Harry was going to make a nice-looking boy one of these days, most people said, when he filled out a little and his skin cleared up. Personally, Harry had his doubts.

So he took a certain discreet solace from the spectacle of his neighbors' humiliation: at least *he* had sense enough to recognize his own privation, at least he knew class when he saw it. The daily appearances of the bus driver, who was, to Harry's mind, the Action's harbinger as well as its pursuer, lent credence to Harry's tantalizing foreglimpse, gave substance to his hopes; and anyhow, Harry liked the driver's *style*.

As it happened, though, the Action came to Harry before he made his way to it; and, irony of ironies, that same bus driver was the one who bore it there, and abandoned it like an orphaned ogre on Needmore's doorstep. Because one stifling dog-days afternoon, as the din of the departing Greyhound died to a mumble in their ears, the loafers looked out through Marvin Conklin's plate-glass window and saw that they'd been delivered something they weren't sure they'd ordered. It stood six feet five inches tall, and weighed in at 238 pounds, and—Harry Eastep saw even while it was still shrouded in a pale swirl of evil blue exhaust fumes, like some unholy apparition the town had conjured up out of the abyss of its own ignorance—it was ugly as sin, and (unlike Harry) it plainly knew no fear, nor any shame.

It was fifteen years of age, and its name was Monk McHorning.

ONE

Harry didn't realize it till later, but he had seen this Monk McHorning once before—years ago, on the front page of the Cincinnati *Enquirer*. The picture had borne the caption *"A Brave Boy Who Wanted to Do A Man's Job,"* and was accompanied by a brief account of how a certain Monroe McHorning, a twelve-year-old at the Perpetual Hope Foursquare Evangelical Reformed Baptist Orphans' Home and School in Newport, Kentucky, had slipped away from the Home and almost succeeded in enlisting in the United States Army.

It was three years later when Harry saw the photograph again, a few weeks after Monk had come to live in Needmore. "Here you go, Step," Monk had said as he extracted the bedraggled clipping from his billfold. "Here's you a pitcher of King Kong from Hong Kong playin' ping-pong wit' his ding-dong." In the photo, a prideful, incorrigibly insolent grin decorated the boy's heavy, already stubble-shadowed jaw, and perched atop his beetled brow was a U.S. Army garrison cap, evidently belonging to the burly recruiting sergeant who stood, hatless and beaming affectionately, at parade rest beside the would-be warrior. The sergeant was the taller of the two by no more than half a head, and on Monk his hat was too small by several sizes. Harry remembered the picture the instant he laid eyes on it. When he'd first come across it in the *Enquirer*, he'd been twelve years old himself, with a rotund little body as bald as a jelly bean, and he'd been amazed, and disheartened, to find that some kid his age had passed an army physical.

Monk's picture was to appear twice more in the *Enquirer*, once on the sports page a couple of months after he'd turned up in Needmore, then not again until more than twenty years later, when it illuminated his obituary.

In the sports-page photograph, Monk wore the uniform of the Burdock County High School basketball team, the Bulldogs—or, as they were variously known around Needmore, with as much despair as disparagement, the Shit-heels, the Fart-hooks, and the Turd-knockers. The picture, cropped to show just that great, grinning head and a span of naked shoulder, was buried deep in a long double cloumn of items headed *"Northern Kentucky Roundball Round-Up,"* Monk's disembodied smirk creeping across the page like some crawly thing the *Enquirer* had netted by mistake in the tight webwork of newsprint. Because most of the merchants in Needmore proudly posted the clipping in their show windows, for several weeks one found oneself confronted with it everywhere—and although in his heart of hearts Harry nurtured an abiding fox-and-grapes distaste for all things athletic, he aspired nonetheless to become a sportswriter someday, and he was an avid student of the sports page. He read the item countless times:

> NEEDMORE—Things are looking up these days for the Burdock County Bulldogs, reports Coach Norbert P. Stickler.
>
> The big, big reason for all the optimism is Monroe "Monk" McHorning, a 6′5″ 240-pound sophomore center who's been sizzling the twines in fall practice, giving Bulldog fans more to cheer about than they've had in many a season.
>
> "I myself personally," Coach Stickler enthuses, "have been working with the boy on his basic essential fundamentals of the game."
>
> McHorning, who played junior-high ball last

year for the Perpetual Hope School in Newport, has recently become Coach and Mrs. Stickler's foster son. Coach Stickler is also the principal of Burdock County High.

"The Good Book tells us a man stands the tallest when he stoops over to help a boy," shrugs Coach Stickler humbly. "I guess that's just the kind of hairpin Norbert P. Stickler is."

The *Enquirer* story didn't say which religion it was whose Good Book Nobby Stickler did business with, nor did it explain precisely how a five-foot six-inch man went about stooping to help a six-foot five-inch boy, nor did it pursue a number of other salient facts and interesting sidelights which were, nonetheless, common knowledge around Needmore.

It didn't mention, for instance, that Nobby Stickler would happily have enrolled and suited up a rabid orangutan if it could rebound. Or that Nobby (himself, personally) had organized a Bulldog Boosters fund, subscribed to at five dollars a month by a number of civic-spirited local merchants, tradesmen, and politicians, for the express purpose of reimbursing himself (personally) for the care and feeding of his prodigious ward—plus, of course, a little loose change for the likely lad to jingle in his jeans down at Craycraft's Billiards on a Saturday night. "We got culture advantages right here in Needmore," Nobby would remind the Boosters as he put the arm on them, "which a orphan boy like him couldn't get exposed to in a million years up at the Prepetural Hope." Or that Monk McHorning played basketball approximately the way Pretty Boy Floyd played cops and robbers, that he threw himself into the game with such violent abandon, felling friend and foe alike with such indiscriminate zeal, that already, after only three weeks of fall practice, half his teammates suffered from various bruises and contusions, that indeed, if the *Enquirer* were to put the matter bluntly, Monk's style of play would have to be described as not merely dirty but psychotic. In Monk's own words, "Scrimmagin' against you feebs is like jerkin' off to a automobile wreck." Or that, not long before Monk underwent the metamorphosis from orphaned waif to foster-scion of the House of Stickler, there had transpired an unfortunate little altercation between Monk and the portly young Foursquare Evangelical Reformed Baptist clergyman who served the orphanage as Superintendent, in the course of which the Superintendent was belabored rather severely about the head and shoulders ("Sumbitch tried t' take my goddamn cigarettes!" the way Monk indignantly told it later. "Out inna goddamn schoolyard! So I was forced t' bust him inna goddamn snot-locker!")—that, in fact, the Superintendent, holding a blood-soaked hankie to his injured hooter, had been in the very act of reaching for the phone to seek the approval of the Home's

Board of Directors to order the vicious young punk into the streets, bag and baggage (not, however, without certain misgivings, for the Board, good Four-square Evangelical Reformed Baptists though they most assuredly were, to a man, tended to share Father Flanagan's view that there is no such thing as a bad boy, especially when the mischievous tyke stands six-foot-five and can pivot off either foot, and when the Home's own varsity had been so woefully inept last season . . .)

—was actually *reaching for the phone* when, almost as though by divine intervention, the Superintendent chanced to recall that his own wife's Aunt Normajean's husband, Norbert, who coached basketball at the county high school in some little burg down in Burdock County, and whom, were the Superintendent at all inclined to be uncharitable, he'd have to describe as a big-mouthed, sawed-off little know-it-all, had persevered, Uncle Norbert had, through umpty-ump losing seasons without respite, and that poor Uncle Norbert and poor, dear Aunt Normajean had neither chick nor child to console them in their trouble, and that—who knows?—maybe the answer to those old dears' prayers was even at that very moment right outside the Superintendent's door, lolling on a couch in the outer office reading a funnybook, with a cigarette butt behind his ear and his muddy feet resting on the needlepoint cushion of one of the Superintendent's favorite antique chairs. Surely the Board couldn't fault him for finding the little scamp a good Christian home, when the lad so obviously needed a father's guiding hand . . .

Anyhow, not a word about all that in the *Enquirer* story; but then when you come to think about it, real life largely consists of things that never make the papers.

Oddly enough, Monk's third, presumably his final, appearance in the *Enquirer*, almost a quarter of a century later, was in many respects not unlike his first. This photo, which graced the front page one Sunday in the early 1970's, was a formal studio portrait of him in full master sergeant's braid and brass, the same old Monk despite the several pounds of suet he'd taken on about the jowls, the same small, quick eyes, beady as ever in the shadow of his garrison cap (this one must have been as big as a peach basket, for it appeared to fit him perfectly), the same old poolroom pallor, the same old malignant curl to his upper lip, the only man alive who could scowl and sneer and smile at the birdie at the same time.

Alive no longer. Because this time the picture was captioned simply "*Slain in Vietnam,*" and it accompanied a story which recounted how M/Sgt. Monroe McHorning, age thirty-nine, "formerly of Newport, Kentucky," a veteran of over twenty years' loyal service to his nation and his flag, had been sleeping peacefully in his barracks in Da Nang when person or persons unknown rolled a fragmentation grenade under his bunk and blew him to smithereens, a devastatingly effective tactic by which disgruntled soldiers in that unhappy conflict sometimes unshackled themselves from the bondage of the Chain of Command. Harry's late mother, then a resident of the Bide-A-Wee Rest Home in Dayton, Ohio, spotted the piece in the *Enquirer* and mailed it to him, along with a note inquiring whether this wasn't "one of your little friends you used to play with in the olden times down in Needmore."

For no particular reason, the day the clipping came Harry stuck it away between the pages of his desk dictionary, where he still comes across it every now and then. As it happened, he put it in the S's, and since he reliably forgets how to spell "separate," he often chances upon Monk's unlovely countenance glowering there among the sibilants, and remembers yet another of Nobby Stickler's favorite Biblical quotations, this one from the Gospel According to the Apostle Grantland Rice:

> When the One Great Scorer comes
> to write against your name—
> He marks—not that you won or lost
> —but how you played the game.

TWO

Monk McHorning inclined toward baseness the way water seeks its own level. That, of course, was exactly what Harry liked about him.

Monk had the dirtiest mind and the dirtiest mouth of any man or boy in the entire recorded history of Burdock County, and he was the dirtiest ballplayer, and he knew the most dirty jokes and shot the dirtiest pool, and his billfold was always stuffed with dirty pictures and little dirty comic books and dirty doggerel: "Now I met May in the strawberry patch / And I gave her a quarter just to see her snatch . . ." He took, he gave the world to understand, no shit, offa nobody. He was big and mean and heavy-hung:

"Tip yer hats to the Big Inch, boys!" he'd holler, grabbing his crotch. "He's got a head like a house cat and ribs like a hungry hound!" He could stick a knife or a hatchet in a tree at twenty paces; he could break wind at will and with incredible authority—the Toothless One, he called his nether spokesman when it was in good voice; he had an obscene tattoo; he French-inhaled; he could wolf-whistle between his teeth and say "Bee-e-e-eer!" when he belched; he'd been drunk—once, he said, on "mackerel-snapper wine" he and some other thirsty orphans had appropriated from a nearby Catholic church; he'd even patronized a Newport whorehouse: "When she seen the Big Inch, boys," he modestly averred, "she give it to me free." He was, in a word, the most *accomplished* personage who'd yet come down the pike in all the days of Harry's ladhood, and Harry thanked his lucky stars for sending him to Needmore.

And by the grace of those same lucky stars, Harry was present that steamy August afternoon, upon the occasion of Monk's advent.

It had been a long summer. Harry was fifteen, remember, which meant that he had somehow already survived three years or so of the sexual blind staggers, three years of lurching drunkenly through the darkness of his imagination like a blind dog in a meathouse, in hot pursuit of whatever female image came to hand or mind—nubile pygmy maidens in the *National Geographic*, the headless torsos in the girdle department of the Monkey-Ward catalogue, the Li'l De-Icers in *Smilin' Jack*, the serene, elegantly coiffed ladies of those magazine ads that bore the enigmatic legend *Modess because* . . . He hadn't yet made a play for the Draw-Me girl on the matchbook covers, but he wouldn't have kicked her out of bed.

Sometimes it seemed to Harry that he'd been condemned to stay fifteen forever, to spend the rest of his days in trembling adolescent anticipation while everybody else copulated and got old and died, leaving him the most severe case of arrested development since Dorian Gray. There was a big secret called "carnal knowledge" going around, and everyone except Harry was getting in on it. Dorian Gray had had to learn the hard way that eternal youth is an affliction—but at least Dorian had been old enough to get his damned driver's license before he came down with it.

So, although Harry's appetites were catholic almost to the point of perfect impartiality, he had met with only token success in his quest for sentient female companionship pursuant to fornication. His problem, aside from whatever personal inadequacies he might be saddled with, was that the girls in his class in school were all older than he was, and the younger ones either liked boys in their own classes or went for older men—seniors maybe, but not fifteen-year-old ones. The only girls who might be impressed by his maturity were freshmen, and Harry was pretty sure people would regard him as a pervert if he started hanging around freshmen girls. They probably regarded him as a pervert anyway, Harry reckoned gloomily, and for all he knew, they were right.

There had been one amusing little episode (that's how Harry intended to characterize it when he wrote his memoirs—"an amusing little episode") last year on the bus after the Limestone game, a ten-minute grapple with a girl named Ramona Halfhill. But Ramona lived way out in the wilds of Burdock County, out of Harry's reach as well as his grasp, so he hadn't followed through yet. Besides, Harry had his eye on heftier fare. Like the ant contemplating the lady elephant with conquest on his mind, Harry had his eye on Oodles Ockerman.

He'd spent that summer in a state of barely animated suspension, working weekends as popcorn popper and ticket catcher at the New Artistic Motion Picture Theatre; loafing away the weekdays in Conklin's drugstore or Craycraft's Billiards or lying around the house reading his mother's Samuel Shellabarger novels and pondering all those heaving bosoms that kept presenting themselves to the Captain from Castile; mowing the yard and weeding Miss Lute's flower beds whenever her impatience with him neared the breaking point; sneaking off at least once a day to the old coal shed behind the house for a few minutes of puzzled meditation on the cutaway view of the female reproductive organs he'd bought for fifty cents from a kid named E-Pod Pennister, who'd torn it out of an encyclopedia in the school library. (E-Pod's name was Dope spelled backwards, but he had a sharp eye for the main chance.) Harry kept the picture under an old coal bucket, along with his pack of Camels and his most prized possession, his condom.

Saturday mornings he'd wash and wax his mother's black 1946 Hudson sedan, that instant anachronism, and then drive it round and round the chickenhouse till it was dusty again.

But if that summer ever ended and it somehow got to be Columbus Day, then Harry too would enter the lists, he too would be a force to reckon with wherever nooky was the prize. For on October 12, if he lived that long, he'd get his driver's license, and one night soon thereafter, in the capacious back seat of the Hudson four-door, in the beefy embrace of Nadine "Oodles" Ockerman, carnal knowledge would be his. Come this happiest of Columbus Days, in some uncharted region of Oodles's immensity, he too would discover his America.

THREE

Monk McHorning did not come unheralded. Thanks to Nobby Stickler's gleeful advance advertising, his impending arrival was the subject of endless talk around Needmore for days before he got there—most of it more or less on the order of the exchange Harry overheard one night that week in Craycraft's Billiards:

"Old Nobs thinks this new boy of his'n is gonna be shit on a stick, don't he?" Fudge Hatton was saying to Claude Craycraft. "He says the boy can do the job."

Claude looked pained. "Nobby Stickler," he said, "couldn't pick his damn mother out of a spastic parade. That kid's one of these juveniles or something, ain't he? He'll cut Nobby's damn th'oat for him one of these nights, you see if he don't."

"If he did," Fudge mused, "then it wouldn't make dick to me if he didn't know a basketball from a damn hedge apple."

The consensus around Burdock County was that Nobby Stickler was an irredeemable blowhard, a self-important ignoramus with a face like a sackful of doorknobs and more hair growing in his ears than on his head, a petty tyrant over other people's children, a quick hand with a paddle, a bully, a whiner and a wheedler . . . and the hapless pawn of his wife Normajean and his father-in-law, old C. V. Nockles, president of the People's Bank of Needmore and paterfamilias of the Nockles Insurance Nockleses and perennial iron-fisted chairman of the Burdock County Board of Education—a man profoundly accustomed to getting his way in all things, at whose pleasure Nobby served.

Nobby, in his turn, had installed a whole infestation of Nockleses on the school's payroll. His distinguished faculty included Normajean's maiden aunt Miss Mary Louise Nockles, an English teacher whose speech was graced by such refinements as "between you and I" and "Don't these flowers smell beautifully!" and "Now I want whomever threw that marble to . . ." and who (or whom) professed to be "one of Mr. Shakespeare's biggest fans." Also Normajean's nephew Mr. Willis Nockles Peed, who taught civics and American geography, wherein he discoursed learnedly on the state of affairs in Mishington, Flahrta, West Consin, New Braska, New Vada, New Yark, and Massatoosetts. And Normajean's widowed sister Naomi Nockles Sizemore, who'd supported herself by teaching home economics since Mr. Sizemore abruptly expired of food poisoning after supper one night in 1943. And Marcella Parsons Peed, Willis's wife, who ran the school cafeteria. And Cousin Roscoe Nockles, the custodian. And of course Normajean Nockles Stickler herself, who served "Doctor" Stickler (his degree having been conferred by the University of Normajean) as office secretary, although she couldn't type, couldn't take shorthand, and usually had too great a mouthful of Vienna sausages and saltines to talk intelligibly over the telephone.

Yet Burdock County would have forgiven Nobby everything, and received him into its heart as one of its own, if only he'd been able to produce a winning basketball team. Back before the war—and before Nobby—the Burdock Bulldogs had been a power in their District, and a force to reckon with even in the Region; in 1937 they'd made it to the quarter-finals of the State Tournament. But then the old coach retired, just about the time that C. V. Nockles's daughter Normajean went off to the state teachers' college and came home with a husband, this little squirt named Stickler, who'd been the assistant equipment manager for the college's basketball team . . . and nowadays the only team the Bulldogs dependably beat was the Northern Kentucky School for the Deaf—and hell's fire, said Claude Craycraft, who ran a little hip-pocket bookmaking operation on the side, you couldn't hardly expect people to stand in line to bet against a bunch of deef-'n'-dumbs, could you now?

So, for all the abiding skepticism, good news about the Bulldogs' prospects wouldn't be taken lightly. Certainly their loyal fans—of which there were, considering, a remarkable number—had had little enough to cheer about of late. The most illustrious Bulldog of recent years was one Guy Gibbs, who'd made the starting five despite the fact that his left leg was some three inches shorter than his right, and was subsequently named an Honorable Mention in the Louisville *Courier-Journal*'s Luman D. Grigsby Most Courageous Athlete awards. But now even Guy Gibbs was lost to graduation, so as things shaped up, Nobby would be obliged to choose his lineup from among the likes of Guy's younger brother, Duck Gibbs, who took his name not only from his curious, flat-footed, rumpy waddle, but also from the webbed toes he was born with; Swifty Grissim, who had fair speed and a pretty good eye and played defense with the sort of insanely single-minded ferocity that kept him right up in the enemy's face in the most maddening way, but who stood only five-four in his double-thick, crepe-soled Flagg Brothers blue suede shoes; Clarence Pennister, E-Pod's older brother, a six-three mouth-breather who boasted a nice, delicate hook shot but suffered from asthma and borderline anemia and weighed 127 pallid pounds, and too often ate his hook before he could get it launched; also a wondrously awkward boy named Foots Hackberry and a piano-legged, dull-witted boy named Norval Stroud and an albino boy named Pinkeye Botts and an assortment of fat boys and walleyed boys and knock-kneed boys and cross-eyed boys and pigeon-toed boys, and, way down at the far end of the bench, the last of the Gibbses, the twins Lester and Chester—better known, respectively, as Six-Finger Gibbs and Four-Finger Gibbs, Lester having somehow acquired, during their joint tenancy of Opal Gibbs's capacious womb, both of Chester's thumbs. Anyway, the lame, the halt, and the blind: that was the peerless Shitheels for you. Last season they'd lost seventeen while winning two, and they owed both victories to the deef-'n'-dumbs.

Still, when Nobby held forth in Conklin's drugstore about his hot new prospect, his audience of soda-fountain loafers was attentive to his ravings. "Now the kind of hairpin I am, myself," he'd been yammering grandiloquently before any little conclave he could muster, "is the thing that pees Norbert P. Stickler off *personally* is a bunch of boys which won't put out for him, like some of these bozo outfits I've had to work with the last few years. But this big scrapper I'm bringing in to play my pivot for me, this McHorning, he's the finest kind of a boy—orphan boy, had a real tough time of it, but y'know Normajean and myself are taking the boy right into our own home to raise, which personally, I myself have always been a great believer that a nice home envirement among a better class of people can do wonders for a boy which hasn't had the advantages—but he's the finest kinda boy, stout as a mule and really cleans them boards . . ."

On the afternoon that this paragon was due in on the Greyhound, Harry wasn't surprised to find E-Pod Pennister and three or four other small-fry occupying the bench on the sun-blasted sidewalk in front of Conklin's, and, inside, a larger than usual colloquy of loafers. There was a palpable hint of expectancy in the air; everybody in the place, Harry realized as Marvin Conklin fixed his fountain Coke, was keeping an eye on the front of the store. Although it was almost bus time, Nobby still hadn't arrived to attend the disembarkation of his prodigy.

Harry was there in part in his capacity as sports editor of the *Bulldog's Bark*, a four-page mimeographed embarrassment that somehow materialized every six weeks or so during the school year. But he was also pursuing certain personal interests. As soon as Marvin's back was turned, he gravitated stealthily toward the rear of the store, and the magazine rack. "NO READING MAGAZEINS THIS MEANS YOU!!!" advised the hand-lettered sign above the comic books. Harry expertly twirled the revolving rack of paperbacks till he located *God's Little Acre,* his current favorite. The book had been on Marvin's rack only a few weeks, but already it was shopworn and dog-eared, and when Harry picked it up, it obediently fell open to a scene in which several hill-billies were displaying some surprisingly sophisticated proclivities.

When the bus came in, he was so engrossed in his study of rural mores that he looked up barely in time to catch the bus driver's disappearing Chessy-cat grin at the screen door, after he'd cast his aspersions and his newspapers into the loafers' midst. As the bus pulled out again, Marvin Conklin was still glowering after him, muttering, "Why, that smart-aleck son of a bee, who does he think . . ." when an audible little gasp of wonder arose from the crowd. A pregnant pause, then Marvin murmured reverently, "Lord help us, what a whopper!"

Harry hastily returned *God's Little Acre* to its niche and made his way to the front of the store, but before he could get a look outside, someone behind him chided, "Hey, boy, you been drinkin' muddy water?" so he went on out the door. And just as he did, the newcomer, who was standing out in the middle of the street with the last thin blue shreds of exhaust fumes still aswirl about him, picked up his suitcase and lumbered over to the curb and stepped up on the sidewalk in front of E-Pod Pennister and his welcoming committee, looked all around him in a leisurely yet purposeful fashion, then said, aloud but to no one in particular . . .

No, wait; first take a good, long look at *him,* and get it over with, for in all candor he wasn't easy on the eyes.

Looming over Harry there on Conklin's sidewalk was surely the most awesome presence who'd ever graced those humble precincts—at least since King Kong and Mighty Joe Young had played the midnight double feature at the New Artistic last Halloween. A whopper, all right; this hulk could have thrown a pipsqueak the size of Harry over his shoulder by the tit, like King Kong tossing away a banana peel. He was broad-assed and barrel-chested, and his shoulders were a yard across, and his head sat atop them like an upside-down coal bucket, broader at the jaw than at the brow. His face was fleshy but very pale except along the stubble-shadowed jawline, and his hair was tarry black, brush-cut on top and greased back on the sides into what, almost a decade later, Harry was to recognize as possibly the original d.a. Behind one ear was parked a half-smoked cigarette butt. His forehead was sloped and shallow, the hairline paralleled scarcely an inch below by an unbroken black band of bushy eyebrow, his close-set eyes a pair of small dark scavengers buried in its shadow. His nose was flat, meaty, the nostrils cavernously splayed, and his jaw was so broad and heavy that Harry half expected it to be hinged, like Frankenstein's monster's, with steel pins at the corners. He wore jeans and a soiled white T-shirt with the sleeves rolled to the shoulders, and at his feet he'd set a road-weary old brown cardboard valise lashed several times around with hairy twine. His biceps were as thick as Harry's thighs, and the whole naked length of his arms bristled with coarse black hair—but there too his coloring was strangely pallid, almost waxen, the shiny bluish pallor of cold lard. Beneath his T-shirt a flaccid pupa of fat ringed his waist. His hands were as big as meat platters, yet curiously white and soft-looking, incongruously delicate, like the hands of some enormous dowager. It would have been impossible to guess his age, for despite his size and obvious strength, despite his whiskered jowls and fearsome aspect, he seemed somehow unformed or newly formed, larval, embryonic yet at the same time faintly morbid, as though he'd spent his days where sunlight never touched him.

"Looks like a damn jailbird, don't he!" somebody whispered behind the screen door.

Harry heard the rustle of a chill wind rising from some clammy, sunless chamber inside his own head, and even in that furnace of an afternoon he shivered slightly when that icy zephyr breathed its message against his inner ear: *There are z-z-zum things, Doktor,* it hissed, *that mortal men were never meant to tamper with.*

So there the alien stood on the sun-struck sidewalk in front of Conklin's drugstore, looking about him in a leisurely yet purposeful fashion, with a decidedly critical eye, his gaze swinging slowly, speculatively, the whole length of Courthouse Street, from Blankenship's Dry Goods to the shoe shop to the Self-Serve Grocery, past Conklin's, past E-Pod and his buddies, and Harry, as if they weren't there, past Hunsicker's Hardware and the marble façade of the People's Bank and the post office and the Snapp brothers' barbershop and Craycraft's Billiards (where Claude Craycraft himself, his wooden triangle yoked around his neck, hung out the doorway staring back at him) and the OK Package Liquors and the White Manor Cafe and Bertha's Beauty Box and the butcher shop and the Farm Bureau Office—stood there with his head reared back as if he owned the place, as if he'd just won the whole worthless establishment on some gigantic penny punchboard, and was considering the various ways he might amuse himself with it, stood there with his head reared back and said, aloud but to no one in particular:

"Well, ain't *this* a hell of a note!"

A long silence, beneath which could be heard the

low mumble of hushed exclamations from inside the drugstore. Then E-Pod found his voice.

"Hey, mister," he said, "are you a jailbird?"

Harry winced, but the stranger only grinned. "Better be a jailbird," he reminded E-Pod mildly, "as a goddamn shitbird."

"Mister," E-Pod said, undaunted, "is that there a tat*tew* on your arm?"

"Take a look, shitbird," the stranger said, "and tell me what you see." He stepped over to the bench and turned the pale, hairless underside of his right forearm to the light and, sure enough, a small blue tattoo clung to it like a spider. Harry moved nearer to join the shitbirds in scrutinizing it—the image of the head and shoulders of a curly-headed woman primping before a hand mirror, all very chaste and proper, the entire affair no bigger than a fifty-cent piece.

"How about you, dad?" He was looking at Harry. "What is it?"

"A woman?" Harry ventured. "Looking in a mirror?" The shitbirds murmured their concurrence.

"Could be somebody's mama," he said, "couldn't it now?" He raised his hand, standing the image on its head, and with the ball of his left thumb he carefully covered the woman's face and the hand that held the mirror, so that just her arms and shoulders were left showing, upside down now to Harry's eyes.

As Harry stared at it, still uncomprehending, the image seemed to slip just slightly out of focus, and when he blinked and looked again the woman's shoulders had somehow become the lower portion of a female torso—and at the crotch the hand that had been innocently primping now played lasciviously in a triangular patch of curly blue hair. The shit-birds crowded around, dumb-struck.

"Any of yez know where Stickler's at?" the stranger inquired, breaking the spell. "Little-bitty fart?"

The description was sufficient. "Yonder he comes now," E-Pod said, as Nobby's bile-green Plymouth, tires squealing, careened around the corner down by Blankenship's. The Plymouth hove up to the curb in front of Conklin's and Nobby hopped out in great disarray, red-faced and sweaty, his feral, hungry little visage locked in a grimace of a welcoming smile, his hand extended for a hearty handshake.

"Monroe, big boy!" he was crying. "Afraid I'm a little tardy—as we say in the education bidness—but I had a flat, and . . ."

"Monk's the name, dad," the newcomer said, not unkindly. "How's yer hammer hangin'?" And he picked up his suitcase and thrust it into Nobby's outstretched hand. "Here, chief, put this grip in the car, will ya."

The owner of the suitcase had handled it as lightly

as a lady's purse, but its weight almost took Nobby to the sidewalk. Monk glanced over at Harry and—to Harry's intense gratification—grinned and winked.

"Well now, M—Monk," Nobby grunted, staggering, "wouldn't you love to go in the drugstore here and have you a nice cool dish of ice cream? I know there's several of our local persons in there which would enjoy . . ."

"Fergit it," Monk said. "I don't want no goddamn ice cream. If they wanta look at me, sell 'em a ticket."

"Heh-heh," Nobby said, lugging the suitcase off toward the car. "Heh-heh-heh."

Monk started after him, then paused by Harry and peered intently into his face, squinting and gnawing his lip, sizing Harry up as though he were about to arrive at some final determination of his fate.

"You got a name, dad?" he said at last.

Harry had been sure his voice had finished changing months ago, but this time it failed him utterly. "Harry EEE . . ." he heard himself squeaking. He gulped and tried again, hoping desperately that he hadn't suffered some sort of awful relapse into puberty. "Harry Eastep," he managed finally.

"I'm McHorning," the other said. Another long moment, while he ruminated further. "I come in on the Long Dog."

"Right!" Harry told him brightly. And then, not

knowing what else to say, he told him so again: "Right!"

"Damn straight," Monk McHorning growled. Their unanimity on the point was curiously reassuring to Harry. Monk looked down the street again, toward Craycraft's Billiards. "You shoot pool, Stepeasy?"

"A little," Harry owned. In actuality, he shot quite a lot of pool—not well, certainly, but often.

"Well," Monk said, "me and you will shoot some, one of these days." He turned and ambled over to the car. Nobby, having wrestled the suitcase into the back seat, was at the wheel. Monk got in and slammed the door. "You girls shake it easy," he said from the car window. "Don't let yer meat loaf." He turned to Nobby. "Move it, chief. I ain't had my dinner yet."

As the Plymouth pulled away, the shitbirds were all a-twitter, and inside Conklin's the volume of conversational hubbub had risen to a subdued roar. Marvin Conklin poked his head out the screen door.

"Boys," he said, "what was that fella showing you all?"

"It was a tat*tew*!" E-Pod told him fervently. "Of a *wo*man!"

Marvin's eyebrows shot up like window shades. "Was she . . . *newd?*"

"Nossir, Mr. Conklin," Harry put in hurriedly. "It was a picture of . . . of his mother, I think."

Marvin withdrew into the store. "Just a tattoo of his mother," he assured someone inside. "I thought he was a damn orphan," someone else observed.

"Hey, Harry," E-Pod wheedled, sotto voce, when Marvin was out of earshot, "that was too a woman's you-know-what on that big g'riller's arm, wasn't it?"

I dunno, thought Harry. *It was, and then again it wasn't.* Aloud he said, "Well, E-Pod, I'll tell you one thing: it wasn't a man's."

A moment later, as he was hurrying off down the street, E-Pod called, "Hey, Eastep, don't let yer meat loaf!" But Harry scarcely heard him. He was on his way down to the Gulf station to tell Swifty Grissim that Nobby Stickler, in his tireless quest for victory, had taken a fifteen-year-old g'riller to raise —and also that, for fifteen, the g'riller sure was suavay.

FOUR

The fact is that, despite all the years he'd lived in Needmore, Harry—hailing as he did from more exalted climes—still felt himself something of a stranger there, a city mouse taking what had proved to be an endless weekend in the country. He'd recognized Monk McHorning right away as a fellow alien and cosmopolite, and he flattered himself that Monk returned the compliment.

In his heart of hearts, Harry had always been something of a stranger. His parents—Leona Pomeroy Biddle, a Needmore girl whose ambition had carried her, despite the onset of the Great Depression, through the Commonwealth College of Commerce at Louisville to an executive position as supervisor of the night-shift operators at the telephone exchange in Dayton, Ohio, and Benjamin Harrison Eastep, a barber at the Orville Wright Hotel in the same town—met late and married later, and Harry was the sole issue of their union. Harry took it for granted that his arrival constituted a stroke of luck—good luck or bad, as the case might be—for Benny and Leona had continued working after they were married, she on the night shift and he cutting hair all day, and it was years and years before they got very well acquainted even with each other, let alone with the Little Stranger who suddenly and inexplicably manifested himself in their midst one Columbus Day, when Leona was in her late twenties and Benny a contemplative, not to say abstracted, forty-four.

Benny and Leona had stayed on in a small apartment in the Dayton rooming house in which they'd met, an establishment which happened to cater mainly to unmarried and retired schoolteachers, into whose willing hands Leona delivered the weanling Harry for tending, while she went back to work. The pet child of a houseful of gentle, clucking pedagogues, old maids of both sexes, Harry had been educated and cultured to a turn by the time he entered the Wilbur Wright Elementary School; he could read and count a little, he knew how to make his letters and color inside the lines, and he had developed the instincts of a born teacher's pet. He prospered, and advanced apace; but among other children he was, and he remained, a stranger.

When Harry was in the fifth grade at Wilbur Wright, there occurred the passing of Leona's father, Elwood Biddle—"Poodaddy" to the family—a part-time farmer and full-time small-time politician, semi-retired. Poodaddy had been, on occasion, Burdock County's sheriff, deputy sheriff, circuit court clerk, county judge, and jailer; at the time of his death he was waging an aggressive, no-holds-barred campaign against his own first cousin, for president of the Senior Men's Sunday-school class at the Methodist church, because, he explained, if a thing wasn't worth fighting for, then it wasn't worth having at all. Harry and his parents had moved in with Miss Lute, Poodaddy's widow, intending to stay just a few weeks while they helped Miss Lute settle her affairs and dispose of the Biddle homeplace, a tidy, prosperous little farm that fronted on one of the back streets of Needmore. The plan had been that they'd all of them be moving back to Dayton in short order, where Miss Lute could become acquainted with the wonders of the modern world and go to the park every afternoon and visit with all the other miserable old retired people from Kentucky, while Benny and Harry renewed their associations with, respectively, Orville and Wilbur, and Leona took out her realtor's license and used the proceeds from the sale of the homeplace to pick up a couple of ve-e-ery interesting little pieces of commercial real estate she'd had her eye on. Leona was a go-getter, and she was ready to get up and go.

But Miss Lute couldn't feature it. She said she was Burdock County born and bred, by Godfrey, and anybody tried to make her leave she'd knock the waddin' out of them, and if they *was* to make her go—or worse, if they left her there all by herself in that big old house way out in the sticks where she couldn't even visit with the other miserable old retired people right there at home—why, she would just will the homeplace over to the penny-stricken little nigger babies in Africa, which might be heatherns but at least they was cultivated enough they wouldn't have drug some poor crippled-up old soul off to live her last dying days with a bunch of Buckeyes. Not that she was racial prejudice, Miss Lute added, glaring at Benny, but she never *could* stand a bloomin' Buckeye.

The upshot was, they stayed—that is, Leona and Harry stayed. Benny gave farming the old barber college try for a year or so, but he was out of his element. A profoundly fastidious man—"persnickety," in Miss Lute's opinion—and a deep believer in the Power of Good Grooming, he despised the work, the hours, the muck and mire, the sweat, the stink, the very crops themselves for the trouble they were to harvest. His straight razor hadn't drawn a drop of blood in twenty years, but when he plowed he left the land in ruin and desolation; he may have been famous in Dayton, Ohio, for the scrupulous precision of his clientele's sideburns, but for the life of him he couldn't keep his fencerows clean. Accustomed as he was to the sophisticated social intercourse of the Orville Wright Hotel, he could abide neither the solitude of the fields nor the briar-hopper incivilities of his neighbors, who, he correctly suspected, made fun of him behind his back for his Ohio accent. Most galling of all, he couldn't even get himself a decent haircut in Needmore; down at the Snapp brothers' barbershop, he warned Harry, they went after sideburns like Mohicans taking scalps.

More and more, Benny spent his time hiding out in the stripping room of the tobacco barn, sipping Four Roses from an old china teacup and listening to Cincinnati on the radio. In Dayton the war boom was in full swing, and according to Benny's information, the Orville Wright was booked to the rafters every night. All the young barbers had been drafted, and the town was overrun with Air Corps officers from Wright-Patterson Field, eager to pay top dollar for non-regulation haircuts. Benny longed for a whiff of Bay Rum to clear the barnyard from his nostrils.

In the fall of '44 came the break he'd waited for, when Miss Lute's still-discerning eye perceived from her post at the kitchen window ("the watchtower,"

Benny called it) that if the barn didn't get a coat of paint soon it was going to dry up and blow away, and she badgered and browbeat Harry's hapless parent until he said he'd do the job. And then three weeks later she found out from a neighbor that he'd only painted the side of the barn that was visible from the kitchen window.

Miss Lute capitulated. "All right," she said, "all right all right all right, go on, then. And don't let the door hit you in the bee-hind on the way out!" she added bitterly as he packed his bag. In a matter of days, Benny was back in business at his old chair in the Orville Wright, and Miss Lute had sold the farm—but not the house, never the house—and banked the money. Then she made out, but didn't sign, a new will leaving everything she had to the Methodist Missionary Society. "They'll know where the nigger babies is at," she explained. She stashed the unsigned will, with Poodaddy's favorite fountain pen clipped to it, in her knitting bag, and retired more or less permanently to her rocker in the kitchen, to wait out her days—which, it developed, were plentiful—in the only company she could stand for long, that of Pittybiddle, her parakeet.

(Miss Lute's chief preoccupation was teaching Pittybiddle to say the names of all her long-dead relatives and ancestors. At the height of his powers, Pittybiddle could reel off the full names of at least a dozen McAtees and Pomeroys, the lineages from which she descended. "*Cousin Gertrude McAtee!*" he'd shriek from his cage by Miss Lute's chair, as though he were announcing the unexpected arrival of guests unseen by all save Miss Lute and himself. "*Uncle Abner Pomeroy! Ethel McAtee! Momaw and Popaw McAtee! Aunt Harriet Pomeroy!*" Sometimes she'd free him from his cage for a bit of exercise, and he'd flicker madly from room to room, calling their names until the whole house was filled with ghostly presences. There were several occasions—for instance, the time Harry was closeted in the bathroom, engaged in private considerations, and Pittybiddle flitted in through the open transom and screamed, right behind his ear, "*Great-grandma Henrietta McAtee!*"—when Harry could happily have wrung the little yellow buzzard's neck and flung him to the cats.)

Meanwhile, Harry's mother, sorely vexed, took another night job, as an operator, not a supervisor, at the telephone exchange in Limestone, seventeen miles away, and she and Harry settled in with Miss Lute for . . . the Duration. Whenever Leona wistfully mentioned Dayton, Ohio, Miss Lute would start rummaging menacingly through her knitting. Grimly, Leona bit her tongue and bided her time.

Every other Saturday, Benny would come down on the Long Dog and spend the night. Leona, who aspired above all else to be what she called a "modernistic" person, kept on the topmost shelf of the kitchen cabinet the makings of something called a Manhattan cocktail, which she said all the modernistic, progressive folks in Dayton, Ohio, were "having" these days, and which she and Harry's father would "have" two apiece of on Saturday evening before "dinner," while across the kitchen Miss Lute fumed and snarled and gnashed her false teeth and furiously careened in place in her rocking chair, muttering about little snips that put on all kinds of Buckeye airs when they didn't even know the bloomin' difference between dinner and supper. Sometimes, to Miss Lute's almost apoplectic consternation, Benny would give Harry the Manhattan-soaked maraschino cherry from the bottom of his glass.

On Sunday morning, while Leona took Miss Lute to church, Benny would set Harry on a tall stool in the kitchen and give him a haircut and a lot of sound fatherly advice about how one ought never to wear brown and blue together, or how the cuff of the well-tailored pant will break just across the shoelaces, or how the real *point*—Benny felt very strongly about this—the real *point* of a good haircut was that it oughtn't to look like you'd had a haircut at all. After Benny was gone on Sunday afternoon it sometimes seemed to Harry he was as evanescent as one of his own perfect haircuts: it was almost as if he'd never been there.

One Sunday, when he discovered Harry was finally

man enough to need a little razor work around the sideburns, Benny endeavored to counsel Harry on matters of a more personal nature. "Son," he murmured solemnly, as his scissors danced *snick-snick-snick* about Harry's ears like a metallic butterfly, "did you know the Bible says 'tis better to cast thy seed upon the belly of a hoor than on the ground?"

"It *is?*" Harry said delightedly, before he could stop himself. He'd always suspected it would indeed be a great deal better, but he'd certainly never supposed the Bible would support him in that opinion. His father, recognizing the unexpected drift their man-to-man had somehow taken, hastily attempted to put it back on course. "Well now," he said, "I mean well now, I mean I didn't mean the Bible means . . . I mean that don't mean it means . . ." He broke off in a fit of coughing so artificial he might just as well have stood there saying "Cough, cough, cough," and when the seizure ended, he cleared his throat several times ("Ahem, ahem, ahem") and said at last, "Also, son, *never* whittle toward yourself."

Harry's mother's method of advising him on such delicate concerns was even less direct, though perhaps more modernistic, than his father's. About the time he turned fifteen, she contrived to place on his nightstand a slim red volume modestly entitled *The Secret of Life (Boys' Edition)*, by Leroy R. Pinckton, M.D. Mostly, the book blithered on and on about pistils and stamens and these mysterious little beasties called spermatozoa, which apparently possessed the remarkable capability of leaping like invisible fleas from misbehaving boys onto unwary girls and fertilizing their eggs for them. But there was one chapter which Harry found endlessly fascinating, and reread countless times.

It described the good Dr. Pinckton's most difficult case, that of an eminently respectable citizen of the community, a certain Mr. Brown (not his real name, Dr. Pinckton was quick to inform his readers), who suddenly and inexplicably threw over his fine Christian wife and his several beautiful children and his lovely home and his prosperous business in order to take up with the town's most notoriously riggish old bawd, a loathsome slattern of evil reputation and foul demeanor. After a year's time, this Mr. Brown came to call upon the doctor, who found him a broken man, a physical and emotional wreck, wasted, hollow-eyed, unshaven and unshorn, trembling all over with little tics and twitches, a mere shadow of his former robust and thriving self. The doctor took one look at his old patient and uttered his prescription in a single word: *Continence!* he thundered; give up this vile woman and her wicked ways, else you must surely perish. At which poor Brown, stricken, blanched and quailed and shuddered piteously; but when at last he'd drawn himself together, he said, rising as if to go, Perish it is then, Doc (or words to that effect), for I shall never leave her. Whereupon the doctor fixed poor Brown with a cold eye across his desk and said grimly, Very well then, go; but tell me first: *Which perversion does she use?* Well, of course the wretched Brown couldn't do it; instead, unable to meet the doctor's eyes, he hung his head and turned up the collar of his threadbare coat and slunk shamefaced from the room, and presumably from the company of decent folk forever.

Harry always suspected that Dr. Pinckton knew the answer to his question before he'd asked it, but the old bastard wasn't telling; enough for him to make the point that *all* perversions lead to ruin. Harry, meanwhile, was left to pore feverishly over the chapter in search of clues—All right, dammit, *which* perversion? What ones *are* there?—as the unfortunate couple wallowed together in his imagination, trying out all the possibilities he could conjure. His poor mother never dreamed how progressive Harry was becoming.

But all the Manhattan cocktails in Manhattan couldn't have made Leona forget that she was stuck in the sticks again, that she might as well have been a million miles from Dayton, Ohio, where the phone company knew a supervisor when they saw one, where "her girls" had said "fie-yuv" and "nie-yun," instead of "fahv" and "nahn" like a bunch of Negroes, and where contempt for hillbillies and briar-hoppers was practically a civic duty. Leona toiled nine hours a night five nights a week at the telephone exchange, until, she said, her eardrums were as sore as corns and her brain was a snarl of bad connections. She got home from work just in time to have a snack and set out breakfast for Harry and Miss Lute; then she went straight to bed, often as not before Harry got up, and slept till time to fix supper and go back to work. It was a relentless, unrewarding grind, but in Leona's opinion the worst of it by far was having to listen to "that old crewd hillbilly brogue" all night long.

So there was Harry, a Buckeye by birth and an urban sophisticate by breeding and inclination, whose encounters with the native population of Needmore and Burdock County during his early years had served mainly to remind him of his aloneness in the constricted little world he seemed fated to inhabit. No doubt, if they'd lived in a house of their own (Harry knows now) instead of moving in with Miss Lute, and if Benny had stayed, Harry would have felt more at home in Needmore, more the citizen, less the mere sojourner. Bookish and plumpish and standoffish, shy as a newt behind his pink-rimmed spectacles, he kept to himself as much as possible those first years, learning what there was to learn in Needmore's schools, struggling to get his Buckeye accent under control, reading Samuel Shellabarger and going to the picture show and, like his father before him, listening to Cincinnati on the radio—force-feeding his already overstuffed imagination.

One of Harry's darker fantasies may reveal something of his state of mind during those solitary times.

He imagined, to begin with, that there was something terribly, terribly wrong with him, some hideous defect or deformity which not only was revoltingly apparent to all who looked upon it but also somehow entailed an equally profound *mental* impairment that rendered him—and him alone—incapable of recognizing his own physical disfigurement, incapable of seeing himself for the repulsive mutation he unquestionably was. When he looked into a mirror, the pudgy, myopic little chap who peered back from the glass wasn't really him at all; the real Harry Eastep was some unspeakable biological montrosity so warped in its so-called mind that it supposed it was more or less like everybody else.

Moreover, in this fantasy Benny wasn't really just a barber in a fleabag hotel in Dayton, Ohio, either. That was merely one of his disguises. Actually, he was a fabulously rich and powerful tycoon, a sort of Daddy Warbucks with sideburns, whose sole preoccupation and concern, aside from administering his enormous business empire, was to protect Harry, his misbegotten spawn, from the awful knowledge of his own loathsomeness. To that end, the great magnate had quietly purchased, right down to the very sidewalks, this little out-of-the-way burg called Needmore, down in darkest northern Kentucky, and driven all its original inhabitants away and replaced them with a huge troupe of actors and actresses, whom he paid handsomely to help him keep his ghastly secret. This of course explained why Harry's fellow citizens didn't start cringing and gagging whenever his vile person hove into view. These were, after all, consummate impostors, real troupers; self-control was their art and calling. Doubtless, they could hold their retching till he was safely out of earshot. *(Orders from the Old Man, see. Keep the kid in the dark about his you-know-what!)* They were all performers in a vast production called *Harry Eastep,* all part of the grand deception, a cast of thousands, everyone who'd touched his life in any way at all—his schoolteachers (he wasn't really smart in school; actually he was severely retarded, but they were paid to give him A's), the editor of the Burdock *Chronicle* (he was rewarded handsomely for including Harry's name in the school honor roll every term), Miss Lute and Pittybiddle (Dame May Whitty and Tweetybird behind their greasepaint), his schoolmates and such friends as he laid claim to, budding little Margaret O'Briens and Mickey Rooneys, nascent stars of stage and screen playing bit parts and walk-ons in a colossal extravaganza for which the only audience was one ill-favored little moron . . . Harry.

As Harry grew older, and his circle of acquaintances increased, the weight of evidence had rendered this doleful whimsy untenable, for the most part. Still, it was something to think about. How much of a bonus had Benny paid Ramona Halfhill to let Harry feel her up that night last winter on the bus? What fee had Oodles Ockerman charged to . . .

Also, things had taken a decided turn for the better lately. About a year ago, he'd awakened one morning to discover that overnight, or so it seemed, he'd grown half a foot taller without gaining a single pound. He went to bed fat and woke up skinny, just like that. Then too, along about the same time, he began to discover that even loneliness has its compensations: in a word, freedom. With Miss Lute effectively out of commission and Leona asleep all day and gone all night, he didn't really have anyone to answer to at home. School had become pretty much of a cakewalk, and what with his gainful employment at the theater . . . suddenly he was a man of means with time on his hands, qualities much admired down at Craycraft's Billiards. And the more he hung out and shot pool and smoked cigarettes and blasphemed and spent his money, the more his fame increased. His status soared; he was becoming, at last and to his considerable amazement and great relief, One of the Guys.

Harry's celebrated intellect had finally begun to pay off too, for it had come to him that the mere fact that *he* had no need to cheat in school did not release him from his moral obligation to help others cheat. In no time at all, his devotion to this principle won him admission to the very highest circles of society: Swifty Grissim owed his eligibility for basketball exclusively to Harry; Foots Hackberry learned more geography copying over Harry's shoulder than he'd ever learned in class; and Ramona Halfhill would probably never have permitted Harry that surreptitious squeeze of her hard, meager little bosom if he hadn't diagrammed sentences for her all year long in Miss Nockles's English class. That was a lot of diagramming, when you considered that the bosom at issue was a tiny affair no bigger than a hickory nut. Harry had searched the whole front of her sweater before he found it.

Ramona, by the way, was a disappointment in more respects than one. A brassy, bony, toothy, knobby hat rack of a girl, she kissed with her mouth wide open, like a snake unhinging its jaws to swallow a frog. Harry had heard about French-kissing, so a couple of times he tentatively stuck out his tongue and waggled it about a little; but when it touched neither sides nor bottom he withdrew it, figuring he must've got the directions wrong somehow. Ramona was a talker, too, and all the way home from Limestone she kept the rest of the bus apprised of the progress of their raptures. "Oooh, icksy!" she'd squeal, as shrill as Pittybiddle. "This little old boy is a-tryin' t'neck on me! This little old boy blowed in my *year!*" But at the end of the trip, as Harry was wiping the steam and sweat off his glasses, Ramona peered at him in the half-light and said, a good deal less stridently, "You know what, Harry? Without them specs, you ain't half bad." Later Harry wished that he'd had the presence of mind to tell her that he understood what she meant, because without his glasses she looked remarkably better herself.

Still . . . she wasn't the Draw-Me girl.

FIVE

Newport! Kentucky's own sinkhole of sin, concrete turf of vice lords and whores and pimps and racketeers, "that stinking Sodom of the Southland" the preachers called it, the fleshpot just forty miles from home where evil lurked in every darkened doorway, where all the world's pleasures could be savored by any man who had the price. Debauched sister city to old-maid Cincinnati, who primly held her nose and disclaimed all kinship whenever a southerly breath of scandal wafted the scent of that prodigal slut's cheap perfume across the Ohio River, Newport strutted her stuff in high-heeled shoes and black net hose along the dimly lighted back streets of

Harry's imagination; took him by the hand and led him down the darkest alleys of desire; wallowed with him amid the fetid rubbish of iniquity and raked his flesh with her painted talons. And Monk McHorning was her bastard.

Until Monk came to Needmore, the most exciting thing that had happened during that interminable summer—with one momentous exception, coming right up—was the time back in July when Harry had gone down to Fudge Hatton's Gulf station, where Swifty Grissim worked, and, lest Columbus Day sneak up and catch him unawares, bought himself a little birthday present in advance: his condom. Fudge kept a carton of twenty-five-cent Supremos, each in its own little pasteboard folder like a matchbook, behind the change tray in the cash drawer. Harry picked a time when Swifty was in the station by himself, and Swifty let him take a whole handful of them into the men's room and try them on, so he'd be sure to get a good fit. They were all pretty much the same, actually, but there was one that did seem to lend the wearer a special dash. The others went back into the box none the worse for wear, as far as Harry could tell.

All of which brings inexorably to mind the one true object of Harry's affections, that ton of fun, the fair but formidable Oodles Ockerman, into whose commodious knickers he aspired to make his way, where ne'er a man had been before him.

Oodles was the daughter of Mr. Newton Ockerman, Jr., Harry's employer at the New Artistic; it was he, in fact, who'd given her the nickname, when she was a thirteen-pound newborn babe. Nowadays, it was generally held around Needmore that Oodles would be a great beauty if she could just lose seventy-five or a hundred pounds. She packed around more poundage just in double chins than most girls of Harry's acquaintance could claim in the places that really counted. People had been telling her for years that she was "just *real* pretty in the face," with the result that she spent hours every day painting and primping and pruning and preening the relatively small area of her geography that had garnered all that acclaim, and mostly left the other two twenty-five to take care of itself—though she did find time to douse it in drugstore cologne now and then, for a rank, floral effluvia hung about her wherever she went, like a slightly sordid idea. Oodles had blanched her hair nearly white with peroxide (it was naturally dark; Harry happened to know that for a fact), and she wore it in an intricately sculptured Madame Du Barry upsweep like a pile of whipped cream on her head, or the droppings of some enormous bird. It was true she had nice, delicate features, but she garnished them so lavishly with powder and paint and rouge that her face looked as though she might've drawn it with crayons on a piece of paper and pasted it to her forehead.

She had big blue eyes as blank as empty teacups, and a tiny red valentine of a mouth, from which issued, in a prating little voice with the remarkable property of instantly reminding the listener of urgent business on the other side of town, an endless word-for-word reprise of all her recent scintillating conversation with her adoring mother:

". . . And I go Well don't be *absurb*, Mommy, you know I woultn't go with no boy which would just set out in the car and blow the horn, and ditn't have the common decentcy to come up on the porch and knock on my door like a gentleman, I don't give doodley-*squat* how Marlene Teeters and Armetta Glasscock and them does, I reckon I am just the old-fashion type of gal, and if a boy was to try and do little Deenie thataway I would just go Hunh-*unh* mister, newp, my mommy ditn't raise me to act trashy, n-o stands for Nadine Ockerman, and that spells nosiree-*bob!* I mean if a boy is supposably such a dern gentleman, then let him be*have* hisself like one! I mean I have got a regular *complex* on that! And so Mommy goes Now Deenie, hon, that type of gal like Armetta and Marlene and them, they can't hep it they was brought up, you know, *vuglar* and all. But I go Oh *foot*, Mommy, you always was one to forgive and forget, and look on the bright side of people, but if you was to hear *one horn* a-honkin' . . ."

So it was not her beauty, nor her sparkling wit,

nor yet her intellect that inspired Harry to yearn so ardently for Oodles's society. Still, his passion had its reasons, both practical and aesthetic.

In the first place, she was nineteen or twenty years old—already in a fair way to become an old maid; and old maids, according to Harry's understanding, were not only dependably easy conquests but were even likely to be, as the well-known poem had it, "grateful as hell." Certainly there was precious little competition for her favors—for that well-bred phantom suitor had never materialized to knock politely at her door, nor had the first horn ever honked. But Harry was not the most self-confident of libertines, and Oodles's very shortcomings gave him heart, and kindled his ardor. He did not entertain the slightest doubt that if and when he ever turned sixteen and got his driver's license and came tooling up before her house in the old Hudson and favored her with a couple of discreet toots on the horn, she'd toss her scruples to the wind and come a-runnin'.

But there was another, even more compelling reason why Oodles Ockerman was Harry's secret heart's desire. It sprang from a little incident in the vacant lot next door to the Ockermans' house.

Several years earlier, when Oodles was just setting up shop as a Lorelei, her father had erected, at his own expense, a basketball goal at one end of the lot, purposing thereby to entice the impressionable youth of Needmore out of Craycraft's Billiards, and possibly into the reach of the vaporous tentacles of Oodles's fragrant influence. At first, whenever there was action in the lot, Mr. Ockerman's corpulent darling would ensconce herself on the front porch and cheer the heroics on the field of play, devouring great snowy hunks of Mrs. Ockerman's angel-food cake and quaffing whole pitchersful of lemonade, while with ecstatic little squeaks and squeals and bursts of applause like wet Chinese firecrackers she made her presence known, just in case Mr. Right chanced to be among the combatants. Any time a post-pubescent athlete chased down a loose ball over near the porch, she was liable to inquire whether he mightn't care to call a time-out and join her in a lemonade and a bite of angel food.

After a while, miffed at unceremonious rejections and outbursts of raucous guffaws and vulgarities from the athletes, she'd flounce back in the house and slam the screen door. Lately, perhaps in the depths of a despair that none of her rude court was man enough to fathom, she'd abandoned her post altogether, and mostly stayed inside when a game was going on.

But if those insensitive louts had been with Harry in Ockermans' lot one night earlier that summer, and seen what he had seen, they might not have been so quick to turn down Oodles's hospitality. For it was on a certain moonless night in June—Oh,

June of Junes! Oh, night of nights!—that he took his usual shortcut through the lot on his way home from Craycraft's Billiards, and as he was starting across the basketball court his eye was caught by some small movement at a lighted window high in the dark wall of the Ockermans' house, and there framed in the window stood Oodles herself in all her monumental naked splendor, as white and dreamily delicious-looking as a marshmallow sundae the size of a haystack, a true abundance of womanly charms, a plenitude, a muchness. She was staring out into the darkness as though she heard the strains of some ghostly lover's serenade, and as Harry watched she raised one hand to her tummy and began idly to explore the wide, pale angel-food hemisphere of flesh the hand discovered there. Half hypnotized, Harry lifted his arms to her and moved like a sleepwalker across the lot toward the house. As he drew nearer he saw the moving hand take up a whole fistful of . . . Oodles, navel and all, work it gently, then lovingly pat the dimpled flesh back into place and wander lower, scratching and kneading, lingering briefly in the shadowy triangle below her belly —did he really hear the *scritch-scritch-scritch* of fingernails attending to some private pubic itch, or has his memory invented that particular?—then wandering aimlessly crab-like up across liver and lights and sweetbreads to settle where the great breasts lay softly splayed upon her chest like half-inflated water wings, the hand tenderly cupping, lifting one breast to offer up a nipple as big and brown and puckery as a coconut macaroon, as though she were entreating the very night itself to come and while away an hour in dalliance with her.

A moment fairly gravid with possibility, Oodles's ample charms for once in her life truly appreciated and she didn't even know it, Oodles Ockerman with exactly what she longed for, a swain so smitten with her form divine that he was actually staggering about in the lovelorn darkness beneath her bedroom window, and she didn't even know she had him.

Then as Harry moved toward the house the basketball goal rose up between them like a gallows, and by the time he'd altered his course enough to see around it, the window was dark, the vision lost. Transported, he hurried home on feet that scarcely touched the ground, as though he glided through the night on invisible roller skates. And all night long in fitful dreams he rooted and reveled in all the nooks and crannies of Oodles's voluminous person, and visited places hitherto unknown to man or beast.

SIX

Harry's mother didn't approve of Craycraft's Billiards, but her main objection wasn't so much to the Lower Element (she called it) that frequented the place, nor even to the amount of drinking that got accomplished there, as it was more expressly to the fact that beer was the chief staple of the adult clientele's diet—for Leona maintained that the surest distinction between a genteel social drinker and a slavering dipsomaniac was in whether or not the suspect took his pleasure directly from the bottle. If Claude had served Manhattan cocktails, she might not have held his establishment in such low esteem.

But her opinion didn't cut much ice with Harry.

He was pulling down eight dollars a week at the New Artistic, and he figured if he chose to invest it in preparing himself for a back-up career as a pool shark, that was strictly his affair, because who could tell when hard times might befall the sportswriting profession? And didn't he have, in Monk McHorning, a unique opportunity to study the fine points of the game under a true master?

Monk shot a mean stick. They played nine-ball mostly, Monk and Swifty and Harry and whoever happened along, a dime a stick, winner take all; and the odds were that, any time Monk got an open shot at the five or six, the game was as good as over. Now and then he'd even drop something on the break and then run the table, the entire game lasting three or four minutes, Monk stalking the cue ball up and down the green rectangle like a great hulking cat toying with a white mouse, handling his cue stick as lightly as though he might at any moment pause to pick his teeth with it, grinning, his jaw jutting like a cowcatcher, his black, beady little eyes a-glitter. Those times his control of his game was so absolute that he'd often be there waiting with his next shot all lined up by the time the cue ball rolled to a stop at the tip of his stick, as if it had eyes and knew just where to go. Harry's tuition in Professor McHorning's School of Pool was running him a buck or so a night, but it was worth every penny of it just to watch the Master when he was inspired, his lumbering grace as he moved about the table, the cue stick dancing like a tailor's needle in his paws as it did its delicate, almost dainty, work, the whole tableau a perfect study in incongruity, Gargantua at his needlepoint.

His basketball, on the other hand, was rather less fastidious.

Almost every afternoon during those last days of summer, there occurred at Ockerman's lot an extended amorphous ruckus which the astute observer might have recognized as basketball, but which more nearly resembled a sort of self-contained civil disturbance, a tumultuous scuffle enshrouded in a cloud of yellow dust, at the epicenter of which one form loomed larger than the rest—the ponderous, cumbrous form of Monk McHorning. Every few seconds Monk's head and shoulders would break through the top of the dust cloud—not that he was all that terrific a jumper, actually; just that there seemed always to be a convenient body or two at his feet for him to stand on—and from that vantage point he'd launch the basketball toward the backboard so hard and in a trajectory so flat he might have fired it from a cannon. Often as not, it'd carom off the board and rip through the net with such force the cords would pop the air behind it. But when it didn't, when it missed the mark and ricocheted back into the fray, Monk, unperturbed and imperturbable, shaking off lesser bodies like a wet dog shaking water, would once more rise serenely above the confusion to receive it, to pluck it like an apple from the sky and slam it off the board again . . . and again, and again, until the One Great Scorer chalked up another deuce beside his name. Then his luckless opponents would take the ball out of bounds, and Monk would smash it down the throat of whichever poor wretch first tried to shoot it, and the scuffle would resume.

Remarkably, the pattern was affected scarcely at all by the quality of his opposition. Monk would select an assistant—usually Swifty when he was available, because Swifty was both a slick ball-handler and a dedicated and selfless team player, who could be relied upon to feed the ball to Monk's inside game without minding too much that he almost never got off a shot himself—and together they'd take on just about whoever happened by, in just about whatever numbers they happened by in. Twenty points was game, and one time Monk and Swifty mortally skunked what must have been one of the most awesome aggregations of basketeers—that's how Harry would have described it if he'd been writing it up for the *Bulldog's Bark*—one of the most awesome aggregations of basketeers ever assembled in Burdock County, consisting as it did of those two Bulldog stalwarts Duck Gibbs and Clarence Pennister and three legendary Bulldog stars of yesteryear, grown men now. Buster Craycraft, who'd led the Dogs of '38 to their last winning season;

Scudder Wallingford, a burly, scowling, reputed wife-beater who in '41 had punched a referee in the mouth in the Burdock County–Limestone game; and the immortal Guy Gibbs, now the senior bag boy at the Self-Serve Grocery—shut out cold, twenty to nothing.

Grown men, and Monk bowled them over like so many duckpins every time he made his move. "God-*damn,* McHorning!" Scudder groused as he picked himself up off the dirt for the third or fourth time. "You was kinda chargin' a little there, wasn't you?" But Monk, his wit honed to the keenest edge, rose effortlessly to the occasion. "Buck you, fuddy," he said. "Whyn't you go out in yer jack yard and back off?" He was no respecter of persons, Monk wasn't.

But he had drawing power. Every afternoon there'd be at least half a dozen spectators hanging around the lot—elderly gentlemen of leisure taking a break from the eternal euchre game at Craycraft's, politicians taking a break from the rigors of the courthouse, sometimes a couple of welfare cases taking a break from taking a break. Nobby Stickler was there all the time, coaching from the sidelines. "Awright now, Monroe," he couldn't resist contributing on one of the rare occasions when Monk let a rebound escape his grasp, "where were you at on that play?" "I was fornicatin' the canine, dad," Monk told him amiably. "Where in the hell was you?" "Hear that, fellas?" Nobby said, turning to the little knot of onlookers behind him. "Boy calls me Dad. Orphan boy, y'know. Didn't have a pot to pee in. And now he calls me Dad."

Even Oodles was back on her porch again, cheering Monk's exploits rather more lustily, it sometimes seemed to Harry, than circumstances warranted.

As a sportswriter and, by default, as the ranking intellect in his social set, Harry remained above the battle, keeping to the sidelines to think important thoughts and provide the overview. Once in a while, when Swifty had to be at the station and nobody else was handy, Monk would press him into service to put the ball in play for him, but Monk took care of the rest of their offense, and most of the defense too. Harry was not athletically inclined. The reason he'd settled on sportswriting as his calling was simply that he loved the Vernacular of the Game—that rich, mellifluous language in which the scribes of the playing field alone are privileged to express themselves; loved the way that, on the sports page, some stalwart is forever sizzling the twines or knocking the hide off the spheroid or booting the oval or pilfering the sacks; loved the assonance, the alliteration, the sheer mythmaking hyperbole, the splendid excess of it all, the poetry!

So he preferred the sidelines, where he could keep the dust off his glasses and compose his rhapsodic odes in the seedy garret of his mind: "Today, on a dusty, sunbaked sandlot down in Needmore, mountainous Monk McHorning made roundball history by single-handedly sabotaging one of the most awesome aggregations of basketeers ever assembled . . ."

The best time of all, though, came at the very tail ends of those days, the last hours before midnight, when Monk and Harry and a few die-hards would repair from the poolroom across the street to the courthouse steps, for the purpose of telling dirty jokes—rather, of listening to Monk tell dirty jokes. He knew dozens of them, hundreds, *thousands,* maybe: jokes about sheep-humping shepherds and soap-dropping sailors and ball-biting sharks and footprints on the dashboard upside-down, about mooseshit pie and the Old Log Inn and how the orangutan got his name (" 'Cause when he swings through the trees his balls go *orang*atangatangatang!"); jokes about nuns and midgets and fairies and old maids, about whores in church and turds in punch bowls and farts in divers' helmets, about Alben Barkley's antique organ and Mrs. Hadley's used centerpiece and the Petrified Penis of the Polynesian Potentate; jokes about JoJo the Tight-Skinned Boy ("Each and ever time he blinks his eye he jacks hisself off") and Bob Cox (Mrs. Cox, poking her head into the barbershop: "Bob Cox in here?" Barber: "No, ma'am, just shave and a haircut") and Johnny Fuckerfaster and Speedy Gonzales and Needledick the Bug-fucker; dialect jokes ("Mandy, I heered yo husbin died of de diarrear. Is dat so?" "Naw*suh*, Rastus, he never died of de diarrear, he died of de *gon*errear! Mah husbin was a *spo't,* he wasn't no shit-aiss!"); and traveling-salesman jokes ("So the farmer's wife brings him a douche bag, see, and he says I reckon you didn't understand me, lady, I ast you could I borry a monkey wrench. And she says Well, that's what I rinch *my* monkey with"); long, lickerish recitations about Paul Revere ("Now Paul was a mighty man, and strong / With a pecker eighteen inches long. / It hung like the pendulum on an eight-day clock") and Christopher Columbus ("He pleaded with the Spanish queen / To give him ships and cargo. / He said I'll kiss your Spanish ass / If I don't bring back Chicago") and Pisspot Pete ("Out of the forest came Pisspot Pete / With snot on his whiskers and shit on his feet"); and, most fascinating of all—because they were not fictions or fancies but bona fide inside info, the real low-down—stories about what went on behind Newport's bright lights ("Now for your Blue Ribbon Special, see, she takes a piece of blue silk ribbon and ties a knot in it about ever two inches, see, and then she . . .").

There was a kind of genius just in the sheer breadth of his repertoire. These were more than mere jokes and loose talk; they were fables, parables, they covered all knowledge, translated his impious vision of the whole of human history into a language even shitbirds could understand. After his own fashion, Monk was a teacher, the very mentor Harry had been waiting for.

And the eight-page Bible was his textbook. "I ain't much on reading, myself," Monk explained. "These is for the Big Inch." He'd assembled for the Inch's reading pleasure the most extensive library of pocket-sized dirty comic books west of King Farouk, trenchantly priapic exposés of the seamy private lives of such celebrated American couples as Andy Gump and Min, Maggie and Jiggs, Popeye and Olive Oyl, Snuffy Smith and Loweezy, Nancy and Sluggo, Mutt and Jeff. Sometimes the principals from various strips would get together in strange and wonderful new combinations: Li'l Abner and Little Lulu, Tillie the Toiler and Donald Duck, Pete the Tramp and Petunia Pig, here a daisy chain made up of Dagwood and Fritzi Ritz and the Katzenjammer Kids and Dick Tracy and Horace Horsecollar and Lady Plushbottom, there a wild debauch in Moonbeam McSwine's hog-waller. There was even one poignant idyll, entitled "The Honeymooners at Niagara Falls," with a text in passable iambics: "The bride was tossing restlessly. / 'Those noisy Falls,' said she. / 'I cannot sleep! I wonder what / Is getting into me!' "

Then too, there was Monk's other associate, that noisy, noisome fellow, the Toothless One. This nether larynx could articulate, at the boss's pleasure, squeaks and rumbles and burbles and booms, could sigh or mutter or moan or whistle, bleat or bark or buzz. Monk had trained it to comment in his behalf whenever it deemed its opinions worthy of a public airing; for instance, it was liable to deliver itself of an enviable razzberry at anyone else's jokes and squeal with delight at Monk's. Whenever it piped up, Monk would cup his hand to his ear, feigning deafness, and importune, "Whisper again, O Toothless One!" And lo, as if to prove that the performance was no fluke, it would repeat its latest utterance to the letter, reproduce every note to the veriest hemidemisemiquaver. The Toothless One's vocal range was really quite incredible; when it was in good voice, it could run the scale all the way from soprano to basso profundo. It was a real artiste, the Yma Sumac of fundaments, and its virtuosity established beyond question that Monk McHorning was a man of parts.

Harry uncovered Monk's prowess at knife-throwing one evening when they and several others were lolling on the courthouse lawn in front of the World War II Burdock County Servicemen's Honor Roll billboard. In a moment of uncharacteristic tactlessness, Harry leaned his head back against the warpy, scabrous plywood facing of the billboard and closed his eyes while Monk told a joke he'd already heard —the one about the little boy whose mother caught him stealing from the cookie jar and playing with himself, both in the same day.

"So when the old man comes home from work that afternoon," Monk was saying, "she tells him how Junior's been carrying on, see, and how he better do something about it. So he jumps up and runs to the cabinet and gets down the big iron skillet. So when she sees what he's got she says Oh honey, you ain't fixing to hit him with that skillet, are you? So the old man says"—something *thwanked* against the billboard perilously close to Harry's head. He peeled back his eyelids and rolled his eyeballs in the direction of the sound. There at the periphery of his vision, not three inches from his right ear, with the tip of its slender, silvery blade buried deep in the plywood, twanged Clarence Pennister's nine-inch fish knife, which Monk had borrowed a few minutes earlier to pare his toenails with. The curved mother-of-pearl handle was still quivering, the blade still chattering faintly in the wood. The blade's tip pierced the exact center of the faded little red star which indicated that Cpl. So-and-so had been wounded in the Recent Unpleasantness.

"It ain't polite to go to sleep when the Well-Built is talking, dad," Monk reminded him indulgently. He leaned forward and plucked the knife off the board. "So," he resumed, turning his attention back to his toenails, "the old man says, Hit him? Hell no, I ain't gonna hit him, I'm gonna fry the boy some eggs! He can't keep that up on cookies!"

Harry would never have let a little thing like a

vented earlobe come between him and Monk. Inevitably, theirs was a Red Ryder–Little Beaver sort of friendship, but Harry wouldn't have had it any other way, so much did he admire Monk, and envy him his worldliness. And he supposed Monk liked him—because to his amazement, Monk *did* seem to like him, after his own fashion—for having the good sense to think so highly of his betters.

It also helped that circumstances at home allowed Harry to stay out at night till the last dog was hung, when Monk was most in need of his good company. Most nights, by eleven or eleven-thirty the symposium in the courthouse yard would have dwindled to just the two of them, and it was then, in those midnight hours with the rest of the town all sound a-snooze about them, that their brotherhood began to discover itself. When everyone else went home, they suddenly had their mutual solitude in common. Harry still remembers the first time it worked out that way.

"Well, how about you, Stepeasy?" Monk said after the last hangers-on had gone home and left them sitting on the courthouse steps in the deep cool of the evening. "Don't I hear your mommy calling you?"

Harry told him the only one liable to be calling him was Pittybiddle, who didn't require an answer.

"The Nobster says I oughta be in bed by ten," Monk chuckled. "I was forced to set him straight on that." The night light was on inside the courthouse, and in the wan glow that spilled from the tall glass doors Harry saw Monk shake his head and smile almost fondly, as though he were recalling the amusing misbehavior of an errant child.

"What'd you tell him?"

"I told him to go take a shit in the ocean. Here," he said, passing Harry the remains of the cigarette he'd been smoking. "You want butts on this?"

As butts go, this one was generous, a real walking stick. Harry was glad to get it. The nine-ball game had cleaned him out that night, and he hadn't had a smoke for hours. The cigarette was a cork-tipped Kool, the first he'd ever had, and when he took a drag it turned his heart to dry ice. He thought the cork tip inexpressibly elegant.

Monk shook another Kool from his pack, fired it up, and lay back against the steps and began to French-inhale, two little white ribbons of smoke streaming over his upper lip and back into his nostrils. He was, it seemed to Harry, the last word in suavay.

"Well, Stepeasy," he said, "are you gettin' any?"

Harry was so startled to be considered worthy of such an inquiry that he almost said "Any what?" But he recovered sufficiently to say instead, "Uh, not too much." He paused to clear his throat. "Not a whole lot," he added; and then, to his own complete surprise, he blurted, "I never did have any, really."

Instantly he wished he hadn't said it. Anything —even a detailed account of his pursuit and momentary capture of Ramona Halfhill's little acorn of a bosom—would have been better than blundering into the admission that he'd never . . . had any.

But Monk was sympathetic. "I ain't gettin' none my ownself," he commiserated. "I wish to hell we was in Newport. I can get laid in Newport quicker'n you could get you a drink of water in this dump. Ain't they none of these old skagmaws around here that puts out? What about that big honker? She puts out, don't she?"

Harry was experiencing what is known as a sinking feeling. "Big honker?" he said querulously.

"Yeh," Monk said. "She ain't fridgit, is she? Does she do the dirty deed?"

"Oodles?" Harry forced a derisive snicker. "I don't think so."

"Well, I just might be forced to give her a try, one of these nights," Monk allowed, settling himself more comfortably on the steps. "I like a big honker, myself. They don't give you no tonker-bone trouble, your big honkers don't."

It was time, Harry saw, to change the subject. "Listen," he put in hastily, not even pausing to inquire what in the world a tonker-bone might be, "if you were up in Newport right now, where would you go?"

"I'd go to Ray's Cafe," Monk said unequivocally. "They got more pussy at Ray's Cafe than Carter's got Little Liver Pills."

At Ray's Cafe, he went on, they were probably missing him right that very minute. When he was at the Home, he said, he used to spend all day polishing up his pool game in the rec room, and then sneak out at night to the nearest poolroom and pick up enough loose change to blow the Big Inch to a midnight Blue Ribbon Special down at Ray's Cafe. If they had wheels, he said, he could get Harry's back straightened any old time; and Harry said in a couple of months they'd have the wheels, all right. Monk said he was damn glad to hear it, him and the Big Inch was hungry for action, they'd be ready any time he was. First they'd cross the river to Cincinnati, he said, and go to the Gayety Burlesque theater and see Rose LaRose and get all worked up. Then they'd head back over to Newport and Ray's Cafe, where the Big Inch's powerful personal influence might even get both of them a little discount. Then Harry got Monk to tell him all about the Gayety, how he'd been going there since he was thirteen, how the last time he'd sat in the front row and seen Bare Nipple, how at the Gayety there was this one old guy—and here it seemed to Harry that Monk's voice and manner sobered almost to the point of gravity, as if he recognized in the old man some nameless peril in their own great enterprise—this one old guy who always sat in the same seat, way off to one side, whippin' his willie under his hat . . .

It was after one when Monk finally stood up and yawned and stretched and said Well, he reckoned even Norbert P. Stickler would realize by this time that the Well-Built don't go according to nobody's schedule but his own. Harry flicked away the smoldering cork tip of what had lately been his sixth or seventh walking stick of the evening, and began gingerly picking himself up off the cold concrete steps. Despite the warmth of the night air, the concrete had set its chill deep into his bones. He was dog-tired and stiff as a board and as full of menthol as if he'd eaten a jar of VapoRub, yet in the interest of prolonging the occasion to the limit, he asked Monk one more question. How, he inquired, did Monk like living at the Sticklers'?

"Aah," Monk said, "the eats ain't bad. But hey, I guess the Nobster wouldn't like me any more than I like him, if I couldn't put the ball in the hoop. Everybody likes you, Step," he added, turning to go, "if you can put the ball in the hoop."

At the far edge of the pool of light that washed the steps he paused, as if he had forgotten something. "Hey," he said, "I got a message for Kay. Can you handle it for me, pal?"

"Kay?" Harry asked. To his best recollection, he didn't know anybody named Kay.

"Yeah," Monk said. "If you see Kay, tell her I love her."

And he sauntered off into the darkness, singing softly to himself:

> "I got a gal in New York City,
> She's cross-eyed and got one titty.
> She can sing and she can dance,
> She's got a mustache in her pants . . ."

SEVEN

Oodles Ockerman was her daddy's girl, all right.

Mr. Newton Ockerman, Jr., proprietor of the New Artistic Motion Picture Theatre, Oodles's father, Harry's employer since the seventh grade, was as ponderous as three hundred pounds of vanilla custard on the hoof, the sort of fat man whose girth was greatest just below the belt, like a gravy boat or a soup tureen. He had the tiniest hands and feet imaginable on a man his size (he bought his shoes in the boys' department at Blankenship's Dry Goods), and a fluty voice and girlish ways. His conversation was punctuated by clucks and smacks and pantywaist expletives—his strongest was "Shithouse mouse!"—

and birdlike flutters of his hands, and he reeked as loudly of aftershave as Oodles reeked of toilet water. Whenever he ventured out on the streets of Needmore there was liable to be some scrawny urchin prancing along behind him with his belly stuck out, aping his walk, that demure, mincing waddle which was symptomatic not of the genital deficiency his detractors suspected but of the bunions and corns and ingrown toenails that resulted from transporting so much mass on such dainty little trotters.

Harry went to work for him as his popcorn boy and ticket catcher on Saturdays and Sundays, the only nights the New Artistic operated, and he soon found out that Junior Ockerman had problems beyond the ken of his podiatrist. Quite apart from the dreadful state of his own health ("I got a bad heart and bad glands and bad nerves and a bad kidney and bad feet," Harry once heard him tell Marvin Conklin, as he studied the shelves of patent medicines in the drugstore. "You got anything for that, Marvin?") he was all a-tremble with anxieties about Mrs. Ockerman's failing health ("She's full of the sugar diabeaties, Harry," he lamented on Harry's first day on the job. "And me complaining about sore feet. Shithouse mouse, Harry, we don't know when we're well off, do we?"); about who'd take care of Oodles when he and Mr. Ockerman were gone ("She's a lovely girl, Harry, just a lovely girl. But shoot, you know the child wouldn't take a job in a pie factory!");

about the Needmore preachers who railed against movies in general and movies on Sunday in particular ("These old backwards preachers can't get their nose out of the Good Book long enough to pay no attention whatsoever to the Finer Things!"); about what was in some ways the unkindest cut of all, the hordes of shrieking children who descended on the theater every Saturday night, ravaging and despoiling, carving their initials in the leather seats and peeing on the floor and dropping condomsful of water from the balcony and blowing their noses on the velveteen wall hangings, running footraces up and down the aisles and filling the darkness with a blizzard of wadded popcorn bags whenever Johnny Mack Brown laid low another rustler. He had an even greater load on his mind than on his feet, poor Junior Ockerman did, and after Harry got to know him he was truly sorry that in his own small way he'd ever been a party to adding to his burdens.

Still, there was no denying that by local standards Mr. Ockerman's sensibilities were pretty delicate. Himself an artiste—he painted flowers on china teacups—he set great store by all forms and manifestations of Sensitivity. He once mentioned to Harry that his main reason for hiring him was that Harry wore glasses, which he took to be incontrovertible evidence of sensitivity in a boy, having worn them himself all his life. Circumstances obliged him to show *Six-Gun Jamboree* and *Tarzan Meets the Bow-*

ery Boys on Saturday night, but for his Sunday programs he favored weepers like *How Green Was My Valley* and *Mrs. Miniver* and *Leave Her to Heaven* and *Mrs. Wiggs of the Cabbage Patch*, and musicals, musicals, musicals—Nelson Eddy and Jeanette MacDonald, Fred Astaire and Ginger Rogers, Lauritz Melchior, Doris Day, Dick Haymes, Jane Powell, Gene Kelly, June Allyson, musicals representing every area of human endeavor, musicals set on the boulevards of Paris, the sidewalks of New York, the snowfields of Saskatchewan, Walt Disney cartoon musicals and Kathryn Grayson–Howard Keel costume musicals and Esther Williams underwater musicals and Sonja Henie musicals-on-ice, musicals about musicals, Mickey Rooney–Deanna Durbin–Judy Garland campus musicals ("I've got it, kids! Let's put on a show!") and leftover wartime USO musicals ("I've got it, Sarge! Let's put on a show!") and Bing Crosby–Barry Fitzgerald faith-an'-begorrah musicals ("I've got it, Father! Let's . . ."). If a movie didn't put a tear in your eye and a lump in your throat, Mr. Ockerman liked to say, at least it ought to put a song in your heart. Personally, his vaunted sensitivity notwithstanding, Harry still preferred the Saturday-night horse opuses, where men were men. Esther Williams was okay, but those twinkle-toed little dudes with pointy shoes and walking sticks gave him the willies.

All the same he loved his job, and he lived for the

weekends, those languid afternoons he spent hovering about the popcorn popper in the vast emptiness of the New Artistic, the evenings when he could lose himself in somebody else's fantasy, somebody else's movie.

The theater stood all alone on a nameless side street just off the courthouse square, where the land sloped away from the curbside sharply enough to have discouraged other construction nearby. It was a frame building, shotgun-style, long and narrow and windowless, flat-roofed, sheathed in brick-patterned tarpaper ("That ol' nigger-brick show-hall!" Miss Lute fumed), otherwise unadorned but for an imposing false front and, above the doors, an illuminated sign that announced the title of the current picture. Inside, the floor of the theater pretty faithfully followed the precipitous contour of the hillside, so that the building was much higher at one end than at the other. Behind the false front there was barely room for the low-ceilinged lobby and, upstairs, the cramped little projection booth, whereas at the far end of the auditorium, where Roy and Gene and Fred and Ginger and Esther and company held forth in Brobdingnagian enormousness, the ceiling was easily three stories high, to accommodate the stage and what Mr. Ockerman liked to call "the largest silver screen in south-central northeastern Kentucky." Mr. O. may have been stretching it there, but the screen was plenty big enough that in the medium close-ups Harry could've tucked himself away like an infant marsupial in the cleavage of Esther Williams's dripping swimsuit.

And of course the aisles were as steep as ski slides. None but the most dispirited child could have resisted the impulse to run down such an incline, and on Saturday night the pounding of little feet often drowned out even the thunder of Trigger's hoofbeats on the screen. There was a balcony (originally for coloreds, before the Yankee dollar lured them all to colder climes), which formed a sort of canopy over the rear third of the great room, and, above that, a celestial expanse of sky-blue ceiling, across which the builder had somehow contrived to paint a dozen or so large, awkward white stars, the kind that children make by starting with the letter A, a little galaxy scrawled across the tin firmament by some home-grown Michelangelo who might just as willingly have crept out upside-down, his dripping paintbrush locked between his teeth, across the very vault of heaven, to put this finishing touch upon his vision.

In its heyday, forty years earlier, when it was the New Artistic Opera House, the theater had been the scene of minstrel shows and popular science demonstrations and musicales and roadshow melodramas, Chautauqua lectures and small animal exhibits and political rallies, puppet shows and talent contests and womanless weddings. Womanless weddings had gone out of style before Harry's time. But his mother still recalled one of those nuptial farces, put on by the Optimists' Club to raise money for Christmas baskets for the needy, in which Poodaddy, a little wisp of a fellow even in his prime, had played the groom opposite a bride portrayed by Newton Ockerman, Sr., the original proprietor of the theater—"a big fat priss just like Junior," Miss Lute put in, with customary charity. Leona had found the whole business mortifyingly hickish at the time, but nowadays her recollection of the long-dead local dignitaries—politicians and morticians and prominent farmers and even a preacher or two—who'd played bewhiskered bridesmaids and fright-wigged flower girls always made her giggle. The best moment, she said, was at the end of the ceremony, when, instead of a kiss, Poodaddy offered his blushing bride a cigar and it exploded in her face.

One time, though, while Leona told the story, Harry allowed himself to wonder what the wedding night might have been like if that strange marriage had been consummated. Even as he smiled secretly at the notion of shrimpy little Poodaddy in his flannel nightshirt and cap, attempting to mount an immense, cigar-smoking prefiguration of Junior Ockerman wearing Kewpie-doll makeup and a dust mop for a wig, he was aware that a disconcertingly similar scene was playing nightly in those sweaty dreams of his, in which he plumbed the murky depths of Oodles's maidenhood. Caught by the short hairs in the clockwork mechanism of history relentlessly repeating itself, he even dared to contemplate the possibility that his lusting after Oodles was incestuous in spirit, that he and she were the monstrous issue of their grandfathers' perverted union, that their bloodlines had been fused ever since those two mismatched old curmudgeons said "I do" on the stage of the New Artistic Opera House.

According to Leona, in those times the New Artistic, with its brilliant footlights and red-carpeted aisles and gold-fringed royal-purple velveteen stage curtain, had seemed the grandest kind of place—the only thing in Needmore, in her opinion, that'd brought it the least bit up to date. But the footlights were long since blown and smashed, the carpet threadbare and black with age, the curtain faded to the liverish color of a deep bruise; when it swept across the stage as the Sons of the Pioneers or Nelson and Jeanette warbled their grand finale, its ratty yellow fringe trailed in the dust like the train of some broken-down old soubrette still trying to play the great lady.

Until he went to work for Mr. Ockerman, the New Artistic had seemed to Harry a creepy, chilling place. He even used to harbor a secret dread of falling asleep during a movie, to be awakened in the dead of night by the resonant, disembodied voices of ghostly old orators, discovering himself locked up all alone in that cavernous darkness with the shades of all those long-gone grandpas and moldy thespians and the unquiet spirits of the legions of redskins who'd bit the dust there. Then, too, there were the rats: during the movies you could hear them, dozens and dozens of sleek gray beasts scuttling dryly beneath the floors and in the walls, restlessly awaiting the moment they could safely venture out to feed on spilled popcorn. When the houselights came on at the end of the movie, Harry had always hastened up the aisle and outside with scarcely a glance behind him, fearful that if he looked back he might see the hungriest and boldest rat come snuffling from its hidey-hole, its ravenous eyes burning fiercely in the shadows.

But after his first few hours alone in the New Artistic his trepidations vanished, and it quickly came to seem the safest, most secure place he'd ever been, his garrison against the mischief of a hostile or indifferent universe. He could sit there huddled dreamily over that warm, glowing, chuckling popper for hours on end, gorged on popcorn, stupefied by the buttery aroma, blissful as an opium smoker, empty of thought, invulnerable to fear, brave as a pioneer at his campfire in the wilderness, oblivious even to the spooks and savages and beasts of prey that prowled the gloom just beyond the reach of the popper's glow and the soft lights of the lobby.

Late in the afternoon on show days, when the popcorn was all popped and bagged and the bags all lined up row on row in cardboard boxes, ready for showtime, and Harry was cleaning up his popper and putting his supplies away, Mr. Ockerman would come bustling in, all harried and undone, apologizing extravagantly for being late ("Lard, Harry, between the doctors and the banks and these old backwards churches . . ."), and they'd turn on the houselights and spend the last hour or so before suppertime sweeping up whatever debris the rats hadn't eaten since the last bacchanal, talking as they worked.

Mr. Ockerman's heft made stooping and bending out of the question for him, so he did the sweeping while Harry moved along before him handling the dustpan and the trash basket and murmuring condolences as, grunting softly at his labors, his employer lamented the iron circumstances of the world.

"Pshew, I tell you, Harry," he'd say, "a refined person don't have a snowball's chance in this town. My poor old mom was the first one to learn me that, and she was right as rain. And she done her level best to help me advance myself and whatnot. But you know her and Dad both passed on the same summer I got my high school degree, so that was the furtherest schooling I ever had. Course it wasn't like now, I mean we got a real education in them days. But when Mom and Dad was gone—they was both quite large too, but delicate some way, and died young for their age—and when they passed on, it wouldn't nothing do me but take over this old show-hall and go into show business. And say, it did just fine for several years, too, we had shows seven nights a week and you couldn't stir 'em with a stick in here. But then the war slowed things down a right smart, and nowadays they've got these daggone radios, and some of them says this old television's quite the coming thing. So you can't tell, you just can't tell. Now Dad left me pretty well fixed, I ain't hurting. But shithouse *mouse*, Harry, I mean most of these sorry yayhoos in this town, if you was to walk up and smack 'em in the face with something uplifting, why, they wouldn't know it from a sockful of poop! And hell's bells, these snotnose young people around here—now I don't mean you, Harry, you're sensitive as the day is long—but some of these snotnoses . . . Like you take the other evening when Nadine—you know Nadine, don't you, Harry? my daughter?—well, the other evening her and this one big old boy was out on the front porch, and I was setting in the living room, and this daggone old boy, he . . . well, he *let* one! A big loud one! Right there in front of her! So blessed loud I could hear it clear inside the house! Never even said excuse me! Why, hell's bells and little shells, Harry, what kinda manners is that? I swear, I'd've smacked the c-r-a-p out of him right there, if he was a boy of mine. But now Nadine,

Harry, she's a very lovely girl, but she hasn't actually had too many beaus, her being large and . . . whatnot, and to tell the truth, I just kinda hated to . . ."

The part of his job that Harry most enjoyed came during the last minutes before showtime, when he replaced Mr. Ockerman in the box office while he raced upstairs to attend to his projectors. At precisely three minutes before eight by the Blackroot Purgative Oil clock on the lobby wall, Mr. O. would begin gingerly backing out of the tiny box office, calling over his shoulder as he maneuvered his Baby Grand of a rear end through the narrow doorway, "Curtain time, Harry! Curtain time!" Hearing that, Harry would quickly lock up his popcorn cashbox and hurry into the box office, stash his cashbox beneath the counter beside the one that held the evening's ticket receipts, take a deep breath, shoot his cuffs, crack his knuckles, and generally ready himself for the most important of all his duties as second-in-command of the staff of the New Artistic.

In the box office there was a fluted metal pipe running down the side wall from the ceiling to about the level of the ticket seller's left shoulder, where it turned out from the wall and flared into a little trumpet of black iron—a speaking tube, its other end upstairs in the projection booth; and just beside it was the main switch that controlled the houselights. Grasping the switch firmly in one hand, Harry would watch the clock tick off the last few seconds before eight, and then he'd put his lips to the cold bell of the tube and say, his voice as solemn as a deacon's, "Are you *ready* up there?" A breathless pause, then the answering voice, tiny as an insect's buzz within the iron flower: *"Ready here!"* And, with that, Harry would throw the switch and plunge the whole house into utter darkness, and the next instant, with a shatteringly brassy clash of introduction music, the great screen would explode to life, the M-G-M lion big as a house and roaring loud as thunder, the Pathé rooster's scream as shrill as a locomotive's whistle, the Republic eagle's wingspread wider than the Great Speckled Bird's.

Sometimes, as Harry perched there with his ear cocked to the cast-iron blossom, waiting for Mr. Ockerman's voice to come skittering like an inspiration down the tube, a great, calm feeling of authority would move within him, as though he were an Old Master seating himself before an empty canvas with yet another masterpiece already springing to his mind. And when with a profoundly satisfying thrust of the wrist he threw his switch, he could almost forget Mr. Ockerman's role in the operation and imagine that he alone commanded those gargantuan shapes and shades to dance their spellbinding dance upon the screen, he alone was at the controls of the dream machine, projecting his own fancies onto the blank consciousness of the ignorant multitudes. This time everybody else was the audience, and Harry Eastep ran the show! And this was no trifling, tinhorn magic-lantern hocus-pocus, this was real *power* he had his hands on, this was . . . art!

EIGHT

In the Burdock High auditorium on the first day of school that fall, Norbert P. Stickler favored the assembled scholars with an address entitled "This Bidness of Libiddy," in which he equated liberty with running in the halls and smoking in the boys' room, and informed them that they had very little of it.

Except, of course, for Monk McHorning.

From the very hour of Nobby's speech—throughout which the Toothless One, to the mixed delight and consternation of those sitting within that worthy's sphere of influence, gave vent to muted little hoots and boos and hisses each time the speaker drove home another point—Monk made it clear that when it came to liberties, the Well-Built would take as many as his delicate nature required. He roamed the halls as footloose as a turnkey, dropping in on his classes like some visiting eminence. He logged countless hours in the boys' room, trading jokes for smokes before an admiring circle of truants and malingerers. And so dependably did he smart off to his teachers and otherwise harass them, so disruptive was his presence in their classes, that his absences were the last thing they'd have complained about.

For that matter, to whom might they have taken their complaint? Was not Monk the boss's putative son, and the last, best hope of the beloved Bulldogs? What lowly pedagogue dast pee into such a wind?

The Toothless One had his own finest hour that fall when Miss Esther Stonecipher, a tubby, hand-wringing little old lady who taught music and also sponsored the Pep Club and the cheerleaders, was attempting to induce a roomful of sophomore boys to sing the school fight song. "All right, boys!" Miss Esther twittered. "Now let's really sing out this time, '*Fight*, Bulldogs, *fight*.' Ready now . . ." and she put her pitch pipe to her lips and tooted it and opened her mouth to sing, but before she could get out the first note, the pitch pipe's toot was answered—"*Phweeee-e-e-eet!*"—by a piercing whistle, exactly on key, which seemed to have issued from the rear corner of the room, where Monk was lolling half in and half out of a seat barely big enough to accommodate the half of him that was in it. Miss Esther hesitated, then tooted again. "*Phweeee-e-e-eet!*" came the reply. Again: "Toot!" "*Phweeee-e-e-eet!*" "Monroe," Miss Esther inquired timorously, "is that you whistling?" "No, ma'am!" Monk assured her. "It wadn't me, Miss Esther! It was the Toothless One!" "The Toothless One? Who's that?" "Why, that's the only name I know for him, Miss Esther! I'm a new boy here, y'know, I ain't acquainted real good yet." "Well, what does he look like, then? There are no toothless people in this room, Monroe." Monk chivalrously declined to point out that Miss Esther's own sandwich clamps were counterfeit. "I don't know what he looks like, Miss Esther. See, I never seen him myself. The thing is, I *can't* see him." "Can't see him? Why not?" "Beats me, Miss Esther." Monk shrugged, the picture of innocence. "I guess I must be nearsighted." With that, a boy named Fits Bigelow laughed himself out of his seat and onto the floor, and Miss Esther, believing Fits to have fallen into one of his celebrated seizures, fled the room.

Even as a scholar, Monk had certain gifts. He had something of a flair for mathematics—not a genius, certainly, but enough aptitude to guarantee him easy B's and C's on the tests. To everyone's surprise, he proved to be a pretty fair speller and one of the few boys in Burdock County High who read without moving his lips—faculties which stood him in good stead with Miss Nockles, the English teacher, although he came perilously close to blowing that advantage the time she overheard him refer to her as "Miss Knockers." (It didn't help at all that Miss Nockles was as gaunt as a hoe handle, and boasted no discernible knockers whatsoever.) He was doing fine in Current Events and Shop—everyone did fine in Current Events and Shop—and in Ancient History, where the teacher, a quavery old artifact named Mr. Birdshit (*né* Burchett), was so myopic he couldn't see past the third row, Monk and two or three other back-row historians were racking up an unprecedented string of matching 100's on the daily true-false quizzes. If he could be induced to write a few themes and book reports for Miss Knockers, and to show up for Civics class and Science and Health class often enough to placate, respectively, Mr. Peed and Mrs. Butts (whose names defied improvement), Monk would be a great success, academic as well as social.

A great success, but not an unqualified one, for he had not found favor with the ladies. His forbidding aspect was part of the problem, of course, as were the madcap pronunciamentos of the Toothless One and the Big Inch's reputation (so far undeserved) for impetuous displays of affection. Also, despite Monk's pallor and his Irish surname, his dark hair and mysterious origins had given rise to speculation that there must be, if not a nigger, then at least a wop, a dago, a guinea—even a Jew—lurking in his woodpile. The nice girls thought him "rough"; to the others he was merely icksy.

Still, it was an auspicious beginning, all in all; yet not so promising that it allayed the last lingering anxieties of the prodigy's self-appointed parent—which is why, one morning during the second week of school, Harry was summoned to that sanctum sanctorum whence all knowledge issued, the principal's office.

"Eastep," Nobby began when Harry stood before his desk, fixing Harry (à la Dr. Pinckton) with a penetrating eye, "Eastep, the Bulldogs *needs* you!"

"*Me*, coach?" Harry burbled. His aversion to all forms of physical exertion notwithstanding, Harry had long kept tucked away among his favorite fantasies several featurettes in which he stumbled upon some miraculous innovation in one or another athletic endeavor—an unhittable secret pitch, a devastating mystery punch, a block-proof hook shot, a knock-'em-dead French kiss—which instantly attained for him his rightful celebrity in the field. But how shrewd and farseeing of Coach Stickler, that unappreciated, much-maligned mastermind of the hardwood, to perceive the genius of Harry's hook shot even before Harry discovered it himself!

"The thing is," continued the Mastermind of the Hardwood, "I need me a good dependable boy like you to be my assistant equipment manager."

Now, Harry's limited experience suggested that

assistant equipment managers were not, as a general rule, wildly successful in the nooky department. Moreover, the head equipment manager was a boy named Shirley Worthington, who was widely held to be what Leona delicately referred to as a "homasectual." Actually, it seemed unlikely that Shirley was subject to powerful libidinous stirrings in any direction whatsoever, deviant or otherwise. A wan, slack-shouldered, whey-faced senior, he appeared to be as sexless as a turnip. And in any event Harry didn't even know for sure what it was that homasectuals did, nor, for that matter, whether he might not do the same things, in a pinch. But he understood instinctively that would-be nooky-masters didn't want to be too closely associated with them in the public eye.

There were mitigating considerations, though: if he took the job, he'd be sprung forever from eighth-period study hall, the last and longest hour of the school day; he'd have a free front-row seat on the bench for all the home games; and when they played away games, he'd . . .

"You'd get to ride on the team bus," Nobby put in, as if he'd read his thoughts. "With the cheerleaders. I believe the little Halfhill girl made the cheerleaders, didn't she?" The man was more cunning than Harry could've guessed. "And you won't have to do a lick of work," he assured Harry. "Shirley Temple tends to that."

Then he went on to tell Harry how the word around school was that he was "smart in his books," and how although he himself personally was a great believer that book sense was no substitute for good plain common horse sense, as an educator he had nothing but the highest respect for the brainier individuals of our society. He'd heard too, he said, that Harry and his big pivot—uh, his boy Monroe—had got to be real buddies lately, and he appreciated that, he really did, because he knew it would mean the world and all to the boy to associate with young persons such as Harry which had been exposed to the good solid educational background all were privileged to take advantage of here in Burdock County. And hadn't he, Norbert P. Stickler himself *personally*, got his start in athaletics as an assistant equipment manager? But he did hope that since Harry's duties, thanks to Shirley, wouldn't trouble him much, he could maybe see his way clear to kind of help his buddy Monroe along, so to speak, with his studies, because it could be a hard blow to the boy's self-confidence if, you know, he didn't make his grades for basketball or something on that order. Like you take Miss Nockles, he said, she had went and got her bowels in a terrible uproar just because Monroe was a few days late handing in his paper on his Most Embarrassing Moment. Of course the boy was just a little shy, was all; but couldn't Harry maybe sort of see his way clear to, uh . . . now he certainly didn't mean Harry was to *write* the paper for him, no indeed, as an educator Norbert P. Stickler wouldn't take a back seat to nobody on being dead set against cheating in any way, shape, form, manner, or means; but maybe Harry could see his way clear to just sit down with the boy and sort of help him along with his, you know, with his words and sentences and paragrafts and what-have-you . . .

"I think he wants me to write it for you," Harry suggested to Monk in the boys' room a few minutes later. Monk hadn't exactly been overcome with gratitude when Harry first reported Nobby's proposal to him. "Help?" he growled. "That little knothead's the one that's gonna need the help!" But Harry's clarification put the matter in a different light.

"Why hell yeah, Step," he said expansively, "you go right ahead! Just pitch right in and say whatever you want to! Hey, I won't even read the goddamn thing!"

"Well," Harry reminded him, "I guess you'd have to copy it over in your own writing."

Monk said he thought he could handle that, as long as Harry made it a short one. "And you better not put in no big words," he added, chuckling. "Get me about a C-plus. I don't want people to think I'm some damn brain like you."

Out of a vestigial sense of obligation to the pursuit of truth, Harry inquired whether there mightn't be some particular moment of embarrassment—if indeed he'd ever suffered one—that he'd like Harry to commit to posterity. Well, he said, sucking his teeth thoughtfully, there was that time up at the Home when he'd been standing in the boys' room taking a leak and some kid had reached in the door and flipped the light switch off, and in the dark Monk had snagged the Big Inch in his zipper and lacerated him pro*fuse*ly about the head and shoulders, and then that night down at Ray's Cafe this one old girl, when she seen what kinda shape the Inch was in, she . . .

But there the memory cracked him up, and when he recovered he and Harry agreed that Miss Knockers probably wasn't ready for that one. "What the hell, Step," Monk said finally, "just go on and make one up. I mean bullshit's bullshit, it don't matter if it's real bullshit or made-up bullshit, does it?" Just then Pinkeye Botts came in and offered Monk a cigarette to show him his tattoo, and after Harry bummed a couple of drags off them he told Monk he'd see what he could do, and went on back to study hall to try his hand at ghostwriting.

Inventing a suitable embarrassment for Monk Mc-Horning turned out to be something like trying to imagine a rooster in jackboots, say, or a hog with a boutonniere. All that day Harry sifted through the small store of information about his subject that he'd accumulated during those long evenings on the courthouse steps, searching for the elusive Idea, one of his jokes or stories or some snippet of his personal history, as Harry understood it, that might pique the interest of the Muse.

The biographical data were pretty sketchy: Monk's first memories, he once told Harry, were of the Orphans' Home. For the first few years of his life there, surrounded as he was by other orphans, questions about his origins just never crossed his mind—and then, by the time he was old enough to entertain the questions, he realized he didn't really care about the answers. "I mean," he said, "whoever left me there sure as hell wadn't coming back to get me. So what did I give a rat's ass who they was?" Since the Home wasn't much on volunteering information, as far as Monk knew—or cared—he might sure enough have been the Son of Kong, offspring of Fay Wray and the King. Simian royalty, princeling of the lower primates.

The rest of Harry's material didn't seem very promising either. Not that Monk wasn't frequently the butt of his own stories: "So ever time I went to give her a big smooch, see, she'd bust out laughin'! So finally I sez Goddammit, Inez, what the hell's so damn funny? And she sez Oh Monkey, you've got one a-*hangin'!* One *what?* I sez. And she sez A *booger!* You've got a booger a-hangin'!" But embarassment implies at least the possibility of shame—and Monk had no more shame about him than a rutting buffalo.

There was one story, though, that Harry thought had promise.

Once—and only once—upon a time, he was telling Harry just a few nights earlier, he had presumed, in a passing moment of weakness, to hope for a piece of the Good Life. It was back when he was only eight or nine years old, on an autumn Sunday afternoon, a time when childless couples often showed up at the Home to window-shop for orphans. Ordinarily, Monk said, he steered clear of those occasions, figuring he was better off where he was. "Lots of them," he explained, "was just looking for some kid to work the ass off of. And there I was, big and stout and all. I used to hide out in the torlet." But on this one particular Sunday, there appeared a young couple who looked like movie stars, the man elegantly turned out in pinstripes and a topcoat and a gray fedora, the woman in a foxfur collar despite the warmth of the afternoon. "Hey, she was a looker, dad! Redhead, see, and had one *hell* of a set of jugs! I seen her, I sez Hubba-hubba, baby, I'll be your little boy any old time!" As the handsome couple strolled about the grounds they were everywhere pursued by a clamorous little crowd of urchins ardently beseeching their smallest attention, and Monk, observing their progress from behind some bushes, found himself in a reverie unlike any he had ever known. In his mind's eye he saw himself the pampered darling of these pampered darlings; he would dwell with them forever in a great house, in a huge sunny room stocked with the most extravagant array of toys; he'd stuff himself with Moon Pies, and never have to snitch his smokes again; his new daddy would pat him on the head and call him "little man" and give him dollar cigars; his new redheaded mommy would clasp him to her helluva set of jugs whenever he required consolation . . .

Suddenly, almost before he even realized he was doing it, Monk had burst forth from his hiding place and was plunging through the crowd of smaller children, crying, in a voice he scarcely recognized as his own, "Hey, lady! Hey, lady!" The next thing he knew he'd bulled his way right up to where she was just rising from bestowing her fickle affection upon some of the more lovable of her small suitors. And there he stood, a full two heads taller than any of the other children and shoulder-high to the lady herself, so that he was eyeball to eyeball with the helluva set of jugs and also with the two grinning, glassy-eyed foxes on her bosom, and he was saying breathlessly, over and over, "Hey, lady, hey, lady!" Then, he told Harry, it was as if a little light went out inside his head, just for an instant, and when the light came on again, whatever it was that he'd been about to say had vanished utterly from his mind, and the one thing he knew for certain was: *Wait a minute, dad, that lady ain't a-bout to adopt no big ugly bastard like you!* And that was when he drew himself up to his full five-two and pointed to the foxes' malevolent little faces and piped, "Hey, lady, them old dead dogs is fixing to bite yez on yer booby!" And turned on his heel and stalked away.

Now there, Harry congratulated himself, was a story he could set to music. He'd need to take certain, ah, libiddies with the facts, no doubt, but his poetic license covered that. Regrettably, the set of jugs would have to go—in fact, he'd better drop the dead dogs altogether—but that still left something to work with . . .

That evening he stayed home from Craycraft's and scribbled and scratched till nearly midnight, and when he was finished he'd assembled a disingenuously ingenuous little essay replete with such small illiteracies as he deemed appropriate to tenth-grade C-plus writers ("I use to be a orphan, I lived in a Orphan Home"), on the subject of how he, Monroe McHorning, a foundling of few years but great size and fierce mien, had once yearned to win the affection of a beautiful redheaded lady who visited the Home ("She must of been real rich, she had a fur coat made out of some foxes on"), and how the lady had been working her way through a swarm of love-starved waifs, dispensing hugs and kisses as profligately as though she were feeding pigeons in the park, and how when she came to him, huge and hulking among his fellows but as hungry for love as the least of them, she took him for the groundskeeper, and shook his hand and complimented the condition of his spirea.

"And that," concluded this doleful narrative, "was the very last time I did ever cry."

Monk copied the paper the next morning in homeroom and turned it in between classes. Miss Nockles, all smiles, handed it back to him that afternoon in English.

"Step," Monk said in the boys' room after class, clapping his amanuensis on the back as he showed him the paper, "you are a regular William Snakeshit, son!"

The pages were gory with red ink, but on the last page Miss Nockles had written:

Whomever could have guessed you were so "deep"!!!

C+

NINE

Fall basketball practice began in earnest during the last week in September and, as was his wont, Monk wasted not a moment in establishing his jurisdiction over the proceedings.

He and the redoubtable Dogs had just suited up and taken the floor for the first time for a few warm-up shots when Nobby Stickler came trotting onto the court tooting a referee's whistle, bawling, "Aw-right, boys, let's do ten or fifteen laps, and then we'll run through some of our basic essential fundamental plays, and then . . ."

"Wrong," his protégé corrected, hauling in a re-bound. "The Well-Built don't run no laps, dad. But, hey," he said, "as soon as I get warmed up good, me and Shortcake here"—he indicated Swifty, then paused momentarily to study the other candidates, who, ignoring Nobby, had already gathered around the new straw boss, awaiting orders—"and him"—Foots Hackberry—"and him"—Duck Gibbs—"and Tarzan over there"—Clarence Pennister—"will take on these other weenies in a little scrimmitch. You go somewheres and blow yer whistle, chief. We'll holler if we need yez."

Six weeks later, when the Bulldogs opened their season against the School for the Deaf, that was the starting lineup: Grissim and Gibbs at guard, Hackberry and Pennister at forward, McHorning at the pivot.

Harry's position as assistant equipment manager turned out to be just as advertised. Shirley Worthington took care of the uniforms, the practice balls, the dirty towels; he ran the scoreboard during scrimmages, swabbed out the locker room after practice, and doctored the players' athlete's foot. Harry, meanwhile, attended exclusively to Monk's eligibility—a commission which (bullshit being bullshit, after all) hadn't proved unduly taxing, and left him lots of free time for lolling in the grandstand with the Bulldog Boosters, who were turning up in ever-increasing numbers to watch the practice sessions.

In the scrimmages, Monk was sensational. It's true that Norval Stroud, the second-string center and big gun of These Other Weenies, played the pivot with all the mobility of a bucket of rocks—so it was no surprise that Monk scored on him at will and inadvertently trampled him to jelly in the process. But there was more to Monk's game than just an appetite for points and carnage. Back in the close quarters of Ockerman's lot, he'd dominated the action by size and strength alone; grace hadn't particularly been a factor. In the gymnasium he revealed that he had *moves* as well. He could cut for the basket off either foot, he could lay down a couple of dribbles when he had to, he could palm the ball like an orange in either of those big white boneless-looking mitts and taunt the defense into committing all manner of ill-advised flailings and flounderings; for laughs, he used to fake poor Norval off his feet by wiggling his ears at him. On the fast breaks he'd move out in no apparent haste, his elbows crooked, hands dangling comically before him, galumphing indolently along like some outrageous upright prehistoric reptile—but by the time the pack reached the far end of the floor he'd be freewheeling like a locomotive, and if Swifty hit him with a pass, woe unto the hapless defender who came between him and the basket. On defense he was invincible, a stone wall with a vicious disposition. Beneath the board he simply used his elbows to describe a circle about himself, within which others trespassed at their peril, while he waited in solitary tranquillity for the rebound. He even proved, in his own way, something of a team player: on no account would he allow the ball to fall into the artless clutches of Duck Gibbs or Foots Hackberry—simple prudence dictated that—and no power on earth could have induced him to employ any play devised by Nobby Stickler; but he willingly indulged Clarence Pennister's hook shot or Swifty's two-handed set shot when conditions seemed favorable, and he and Swifty had worked up several snappy little razzle-dazzle routines between themselves. *That big old hammerheaded orphant boy can do it all, can't he?* the Boosters told each other happily in the grandstand. *You better believe it now*, they said, *this big McHorning come to play!*

Prominent among Monk's admirers was Claude Craycraft, but lately numbered with the skeptics. Claude hadn't given Nobby the time of day back in August when Nobby first offered him charter membership in the Boosters, but he knew talent when he saw it, and now he was a subscriber in good standing. Any time Monk came in the poolroom Claude had a little something for him on the side—a couple of bucks, a free practice round on an unemployed table, a pack of smokes, a free Coke or a Moon Pie or even, now and then, a bottle of beer if Monk would step into the pissery to drink it. After fall practice got under way, Claude was on the scene for almost every session, sitting a little off from the others, his felt fedora pushed to the back of his head, a kitchen match in the corner of his mouth, his muddy, hooded eyes as impassive as a lizard's. But whenever Monk negotiated one of his slicker moves, Claude's mouth would work silently, as if he were doing figures in his head, and the kitchen match would execute a nimble dance along his pendulous lower lip.

Shirley Worthington was Harry's own cross to bear. Not that Harry disliked him—there wasn't a reason in the world for disliking Shirley—but he did tend to be sort of sticky. Shirley's mother was Bertha Worthington, a widow, who operated Bertha's Beauty Box. She was teaching Shirley the rudiments of hairdressing; already a number of the Beauty Box's patrons steadfastly held out for Saturday appointments, when Shirley was available to "do" them. Oodles Ockerman's latest do, a towering, lacquered upsweep that made her look like a rhinoceros with a futuristic horn, was Shirley's creation. He thought he recognized in Harry a kindred spirit, and was always trying to talk to him about "coyffewrs." Harry listened politely, but he didn't encourage the connection. He was afraid it wouldn't do much for his new, hard-won, and probably shaky reputation as a man among men.

Monk and Harry and Monk's other acolytes passed their evenings much the same as ever, except that Monk had lately taken to quitting the nine-ball table as soon as he'd staked himself to a pocketful of other people's dimes, in favor of the check-pool game on the front table, where Scudder Wallingford and Fudge Hatton and Claude Craycraft's son Buster and sometimes Claude himself were usually shooting two-pill check for fifty cents a stick.

Now, two-pill check at that price can be a fairly fast game to start with, and when one or two of the local high rollers took a stick—Rodney Kiddington, the jauntily no-account son and heir of Needmore's only physician and wealthiest citizen, or Ace Fraley, the Standard Oil agent, or Gene Weaver, who traveled as a buyer for Brown & Williamson in the winter and wasted his substance in riotous living the rest of the year, or Pillbox Foxx, the veterinarian, or some of the courthouse crowd from across the street—the competition for spoils was liable to get spirited. A good deal of capital could change hands in short order. It was nothing unusual for either Kidney Rottington or The Weave to donate fifteen or twenty dollars in an evening's play when they were drinking. Ace Fraley was a streak-shooter who might win two or three games in a row and then lose for nights and nights on end. Pillbox was a one-eyed man with a passion for gambling—an unhealthy combination in a poolroom—and the politicians couldn't shoot pool for talking about themselves. Fudge and Bus Craycraft were steady, consistent players who reliably won more than they lost, but Scudder's game suffered from his churlish nature; it burned old

Scudder's mortal soul for somebody else to make a good shot, and when his own turn came he'd slam into the cue ball as if he intended to exact retribution on it for collaborating with the opposition. And Claude, an odds-player to the bone, *always* came out ahead, not because, despite a lifetime spent in poolrooms, he was all that handy with a cue stick, but simply because he'd only buy into the game when either Rodney or The Weave was drunk, or Ace was in a losing streak, or Pillbox had the fever, or Scudder was incensed.

Of course, Harry would never have presumed to join such fast company himself, but he'd usually leave the nine-ball game when Monk did, and take a seat on the pop cooler with the railbirds, where he could observe at his leisure his friend's progress in the big time.

Monk was doing fine. The competition was stiffer, certainly, than any he'd encountered on the nine-ball table; and in check-pool, sheer luck figures almost as importantly as skill, so no one player can ever control the game completely. But, as Harry was discovering, a true shark makes his own luck, to a substantial extent. For instance, whenever Monk was shooting just ahead of Scudder Wallingford in the rotation, he seemed invariably to leave Scudder some absurdly unplayable circumstance to deal with —outrageous leaves, leaves that would have been but little worse had Monk simply dug a hole in the table and buried the cue ball and bricked it over. Eventually, this would put Scudder out of sorts, and set him to muttering—darkly but not too loudly, and then only when Monk was in the pissery—about dirty pool and city-slicker sonsabitches.

But Monk got on very well with the other check-shooters, even when his pockets were bulging like a bandit's saddlebags with their own silver. For these were sporting gentlemen; above all else, they loved a winner. A boy that didn't come down with buck fever in a pool game wouldn't catch it on the basketball court either, and such a boy would surely win some damn ball games for a change. They received him unto the very bosom of their company. Anxiously, they inquired after his health, his grades, his state of mind; they fawned over him, flattered him, fondled him, clapped him on the back, felt admiringly of his muscles, slapped him affectionately on the rump. "If any of them knotheads was women," Monk said privately, "I wouldn't be so damn hard up alla time." He was, they owned, a pip, a prize, a peach, a prince; everything they had was his. For him they'd even turn on one of their own number: one night when Scudder's grumblings got too uncomplimentary to be ignored, and Monk was forced to point out to him that he didn't have enough hair on his ass to make a wig for a grape, the other players hooted and jeered poor Scudder off the premises.

Another night, after closing time, Kidney Rottington and The Weave even included Monk in a foray to Limestone, where they cruised the shadowy back streets of Stringtown, the colored neighborhood, in Rod's old Buick rag-top, drinking beer and searching for a species of wildlife they called "smoked chicken," which, they'd assured Monk, would practically fling itself beneath the Buick's wheels in its desperation to attract the smallest notice of such studly Caucasian personages as themselves. Unhappily, the only specimen they managed to flush that night promptly bounced a rock off the Buick's windshield and hollered for the po-leece. From then on, nonetheless, whenever Rodney and The Weave encountered Monk they'd slap him on the back and call him "that old smoked-chicken man." Monk suffered these intimacies with a certain quiet dignity, and led the perpetrators forthwith to the check-pool table.

Before the cool weather set in, Monk and Harry stopped by Oodles's front porch a time or two—in order, Monk said, to start getting her all hot and bothered. His technique for inflaming her ardor was deceptively simple: he merely pretended to discern powerful lubricious imputations in every word the poor girl uttered. If she remarked that it was nice out, he'd say, Yeh, I think I'll leave it out. If she said she felt a little cranky, he'd point to the Big Inch and say Y'wanna feel a big one? She couldn't use the words "do it" or "come" or "box"—or even "fork"—in any context whatsoever without sending him into mock paroxysms of priapic ecstasy. "She's gettin' hot to trot, Step," he declare afterwards. "When they're hot to trot, they'll drop their jeans for jelly beans."

(Years later, as a graduate student, Harry wrote a transcendentally fatuous seminar paper entitled "Manifest Destiny and the Myth of *Penis Captivus*: A Freudian Interpretation of the Westward Movement." It was three o'clock in the morning, and Harry, giddy with exhaustion, was typing the final page—"Thus we see that the westward thrust was impelled primarily by the age-old Anglo-Saxon neurosis *vagina dentata*, and from the onset it was essentially a quest not merely for new lands but for America's lost manhood"—when he suddenly fell into a fit of rueful giggles, it having just occurred to him that his analytical method owed more to Monk McHorning than it did to Freud.)

Remarkably, Monk's approach seemed to be working like a charm. Oodles received his attentions with a simper, and called him "fresh" and "wolfish"—but elsewhere she was telling everyone that "this new boy is just real different!"

Meanwhile, Harry, an obscure figure skulking and brooding in the background of these proceedings, had begun to entertain misgivings regarding his own expectations. Once Oodles and the Big Inch had been introduced, he reasoned, it wasn't going to be easy to impress her. Nor had it escaped Harry's attention that when Monk had gone off with Rod and The Weave on that fabled smoked-chicken run, he'd been home writing Monk's composition on What He Did Last Summer. It'd serve Monk right, Harry peevishly advised himself, if when he came a-huggin' and a-chalkin' to the summit of Mt. Oodles, he found the flag of Speedy Gonzales Eastep already fluttering gaily on the breeze.

The problem was that, lacking Monk's command of the social graces, he didn't know how to go about getting Oodles all hot and bothered. Whenever he tried larding his conversation with salacious innuendo, he sounded not seductively suavay like Monk but even more deplorably adolescent than he felt. He could try warming her up with a couple of Monk's seamier jokes, he supposed, but somehow they always seemed to lose something in the retelling. (As conversational foreplay, the gambit probably wasn't very promising anyway. Harry's scenario for the liaison went more or less like this: He'd say, "Uh-did-you-hear-the-one-about-this-guy-goes-to-the-army-and-he's-gone-oh-about-two-years-and-one-day-his-little-boy-looks-out-the-window-and-hollers-Mommy-Mommy-here-comes-Daddy-with-a-Purple-Heart-on-and-his-mommy-hollers-I-don't-care-what-color-it-is-lemme-at-it?" Then Oodles would say, "Tee-hee!" and Harry would yelp, "Let's do it!" and pounce upon her, humping like a steam-powered gigolo.) He even considered borrowing a few of

Monk's eight-pagers for the occasion, but he certainly couldn't tell him why he wanted them, and Monk's conjectures would have been discomforting.

But even as he was weighing these unpromising alternatives, Fate was conspiring to present Harry with a livelier possibility.

TEN

All was not as well as it might have been those days at the New Artistic, where attendance had been on the wane since early summer. At first, Mr. Ockerman had attributed the decline to the usual hot-weather doldrums, but by September, when the weather turned cool and the crowds grew smaller still, he'd made a positive identification of the culprit: television.

Hunsicker's Hardware had sold dozens of the infernal devices in the past few months, with back orders for dozens more. Antennas were springing up around the countryside like some giant new noxious weed. Mr. Hunsicker had installed, as an inducement to trade and for the edification of the poorer classes, an immense Sylvania in his display window, with an automatic timer that shut it off at night at a whimsically predetermined hour. The Sylvania towered like a mahogany outhouse, but the picture was scarcely larger than a knothole, and the knot-hole was almost always out of adjustment; and of course the sound couldn't begin to penetrate the plate-glass window. Yet every evening after supper, fair weather or foul, a handful of impecunious dreamers would assemble on Hunsicker's sidewalk and stand rooted there for hours, gravely studying the fleet, flickering miniatures in a kind of cataleptic rapture until, perhaps with his lips moving as if in a tiny scream of protest, an elfin Milton Berle or Herb Shriner or Chief Don Eagle would be abruptly sucked into a brilliant pinprick of white light at the center of the screen, the little dumb show gone like quicksilver down the drain. An instant later the pin-prick would swallow itself as well, and then at last the communicants would turn to one another—to people they'd known all their lives!—blinking their amazement, as though the Sylvania had just gulped down their dearest, most intimate friend, leaving them abandoned among total strangers.

Meanwhile, households that didn't have television yet were discovering a wonderful affinity with nearby households that did; inexorably drawn through the dusky late-summer evenings to the glowing tube, neighbors gathered in the sitting rooms of homes marked by the Sign of the Antenna, to contemplate the Finite. Later, as they stumped home in the dark, they vowed: *I'm gonna have me one of those!* And on the Largest Silver Screen in South-Central Northeastern Kentucky, the great stars played to empty seats.

"But shithouse mouse," sighed Mr. Ockerman, counting his dwindling receipts, "that's show business, Harry."

On the first Saturday afternoon of October, less than a week before Columbus Day, Harry was at the theater just finishing his last popper of popcorn when Mr. O. rushed into the lobby fairly trembling with excitement. "Lookahere, Harry, lookahere!" he called breathlessly, waving a long white envelope. "Some bunch is a-wanting us to put on a show!"

In the envelope was a handbill advertising a movie, produced and directed by one Philander C. Rexroat, entitled *Dads and Mothers* ("Screenplay by Philander C. Rexroat"), purporting to be "The World's Greatest Sex Hygiene Attraction," narrated by Philander C. Rexroat, and accompanied by a lecture ("On stage! In person!") by Philander C. Rexroat, eminent scientist, author, lecturer, scholar, theologian, world traveler, sexologist.

SEE THE BIRTH OF SWEDISH TRIPLETS BEFORE YOUR VERY EYES! the handbill shrieked. SEE THE HORRORS OF VENEREAL DISEASE! SEE THE MIRACLE OF THE HUMAN REPRODUCTIVE SYSTEM! NO ONE UNDER HIGH SCHOOL

AGE ADMITTED! SEPARATE SHOWINGS FOR GENTS AND LADIES! IT MAY SHOCK YOU BUT IT WILL MAKE YOU THINK! SO SHOCKING A REGISTERED NURSE MUST BE IN ATTENDANCE AT PERFORMANCES! A PHILANDER C. REXROAT PRODUCTION!

This text was illuminated by a stiff little drawing of a fully clothed couple—indeed, they were in evening dress—in a violent yet abstracted-looking embrace. The woman bore a vampish resemblance to the one in Monk's tattoo; the gentleman sported a lounge lizard's slicked-down hair and mustache. At the bottom were blank spaces where the name of the theater and the dates, times, and price of admission could be written in. There was also a letter, beginning "Dear Mr. Theater Operator," which announced that the illustrious Dr. Philander C. Rexroat, "the Cecil B. De Mille of Sex Hygiene Entertainment," was presently on tour with his famous film, and that although "this colossal educational attraction" had been playing to packed houses throughout the South, there were still a few bookings available if they acted quickly, Dr. Rexroat to provide the Registered Nurse as well as two hundred copies of the handbill, he and Mr. Theater Operator to split the gate 70–30, the lion's share to go to the Doctor, for the furtherance of his great educational work. The letter advised interested parties to call a certain Cincinnati phone number for information, then concluded: "Reminding you that 'tis better to light one candle than to curse the darkness, I remain, Yours for Better Sex Hygiene Entertainment, Philander C. Rexroat, Sex.D."

"Ain't that us to a tee-total, Harry?" Mr. O. exulted. "I mean about cursing in the dark and whatnot? If that ain't this old town made over!"

Harry didn't respond. He was thinking he might give up sportswriting, and go in for sexology instead. He hadn't realized you could make a career out of it.

"Harry, daggone it, I'm goin' up to the house and give the fella a jingle! Shoot, this'd be just the trick to wake this old place up! Why, shithouse mouse, boy, this here's the Twentieth Sentry!" Mr. Ockerman's enthusiasm had propelled him halfway back out the door when an afterthought drew him up short. "Say, Harry," he said, suddenly apprehensive, "you don't reckon it'd be . . . nasty or anything, do you?"

"Well," Harry reminded him, "he's a doctor and all. They'd get his license, wouldn't they?"

"You bet they would!" Mr. Ockerman beamed. "They'd get his daggone doctoring license! I expect he's one of these doctors of this old vernal disease. If it was nasty, he wouldn't be hooked up with it, would he? I mean, shoot"—he said this on his way out the door again—"I mean, I'd send my own daughter to a show like that!"

Aha! quoth Harry inside his head, as inspiration

lit him like a light bulb. If the Cecil B. De Mille of Sex Hygiene Entertainment couldn't get Oodles hot and bothered for him, then she was stone-cold fridgit sure enough! And if there was an unfortunate irony in Mr. Ockerman's being made the unwitting agent of his daughter's undoing—well, who was Harry to question Providence? Bring on Cecil B. De Mille, the Swedish triplets, the Registered Nurse! Bring on the Horrors of the Human Reproductive System, the Miracle of Vernal Disease!

Bring on . . . the Big Honker!

ELEVEN

Columbus Day. Harry's mother gave him what she'd given him for his previous three birthdays, a subscription to the *Reader's Digest*, her favorite magazine. (One of Leona's fondest fancies was that Harry would someday be inspired to "write up something cute" for the *Digest*'s "Life in These United States" department, and get his writing career off to a hundred-dollar start. She persisted in advancing this scheme even in the face of Harry's assurances that sportswriters eschewed all things cute as a matter of professional conscience.) Granny presented him with her usual hand-knit socks too thick to get his shoes on over, and his father sent his usual nickel birthday card with a dollar bill inside. The card was signed "Your father, Benny."

But Harry got his driver's license, next to which accomplishment these gala festivities necessarily paled. For now his strategy was essentially intact: Dr. Rexroat & Co. were to play the New Artistic on the evening of the first Saturday in November, two performances, ladies-only at seven-thirty and gents-only at nine, and during the second showing, while every other post-pubescent male homo sapiens in Burdock County was held in vicarious thrall by Dr. Rexroat's shadow show, Harry and the decidedly three-dimensional and uncontrollably hot and bothered Oodles Ockerman would repair in the Hudson to some dark lane outside of town, for purposes of a fornicational nature. Thanks to the stimulating effects of Dr. Rexroat's presentation to the ladies, he probably wouldn't even need to notify Oodles in advance; at the first tootle of his horn she'd be ready as a radio.

In the meantime (Harry told himself, girding his loins), he'd have to wage a holding action against certain unprincipled parties who'd employ every wile and guile to relieve an innocent maiden of her uneasy virtue. Vigilance would be his watchword; from now on, he'd stick to Monk as tight as his tattoo.

Leona still used the car six nights a week, but on weekdays Harry had exclusive title to it from sunup to suppertime. Every morning he'd sedately drive *en grand seigneur* the five hundred yards from home to school and park beside the schoolyard, so that at noon he and Monk could ride around and smoke Harry's cigarettes. In the afternoons after basketball practice Monk luxuriated in the shower till five-thirty or so and then spent another leisurely quarter of an hour working his d.a. into shape, while Harry hung around the empty gym waiting to drive him home to supper—another five hundred yards in the opposite direction. Then Harry would rush home and bolt his own supper and beat it back uptown to Craycraft's, to be at the ready when Monk needed him to contribute to his check-pool stake and hold his coat and chalk his cue and otherwise assist him and admire him and appreciate him and bask in his presence until the wee small hours. Next day in study hall he'd wake himself up long enough to do Monk's homework—Monk's grades were improving, but Harry's had slipped into a slow decline—and the next thing he knew, there he'd be in that empty gym again, languishing half-asleep in the bleachers, nodding off to the sweet strains of Monk's rendition of "Yank My Doodle, It's a Dandy" issuing in sonorous atonality from the showers. In short, Harry strove to make himself indispensable, and Monk, true to form, proved not the least bit loath to take advantage of the service.

So unabashedly available was Harry that he even managed to get himself introduced into the society of grown men, if Kidney Rottington and The Weave could be said to approximate that description. As soon as that antic pair discovered he could be relied upon to keep them in smokes and to kick in for gas and beer, they were pleased to include him in their midnight rambles with Monk. And when Harry started letting them in free at the movies (Monk also got free popcorn), why, they were positively honored to associate with a man of so much property and influence. In return, they were teaching him to drink beer (already he could get down a whole bottle, though by the time he got to the last swallow it was always as warm as spit, and almost as hard to drink) and improving his vocabulary and widening his horizons to include, for instance, a taste for smoked chicken (or, more precisely, a consuming interest in cultivating one) and a very definite preference for the company of men to that of boys.

Gene and Rod were wonderfully learned masters for an apprentice reprobate to go into practice with. Seedy but still good-looking young rounders with boyish smiles and frolicsome dispositions, amiably degenerate, they were worthless to a fault. They'd been as thick as thieves since childhood, had gone to high school together, been second-string Bulldogs together, joined the navy together, served together in the South Pacific, survived intact together, and been mustered out on the same day. Then they'd hastened home to Needmore to become charter members of the 52–20 Club, a government program whereby returning combat veterans were paid twenty dollars a week for up to fifty-two weeks, while they readjusted to the rigors of civilian life. Rod and Gene had some serious readjusting to do, in consequence of which they stayed as drunk as owls for the entire fifty-two weeks.

Since then Gene had manfully struggled his way to functional sobriety each winter to serve his four months of penance on the tobacco market, but Kidney Rottington remained perfectly undefiled by honest toil; he lived by sponging off his despairing old daddy, Dr. Kiddington, and by trading and re-trading an endless assortment of decrepit automobiles, in which he also seemed, principally, to dwell, banging about the county like some marvelously expeditious tortoise. Gene was married to a lovely, uncomplaining girl named June, who worked in the Bank, and he frequently extolled the virtues of matrimony to Rod, although Gene himself had experienced scarcely two weeks of married life in the two years he'd been married. The rest of the time he was either on the road with the market or on an extended toot with Rod.

Rod and Gene didn't look at all alike, yet they were so very often together, and so very much alike in so very many other ways, that they almost seemed a single presence, like Siamese twins of the fraternal variety. By way of individual description, suffice it to say that one of them was either tall or short, and either dark or fair, and so was the other one. They even shared a nickname. "See now, Sharpie," Gene might cajole, "if you was a married man, you wouldn't be out drinkin' with ol' sots like me." "Why hell, Sharpie," Rod would point out, "*you're* married, and *you're* out drinkin'!" " 'At's right, Sharpie," Gene would have to admit, "but I wouldn't be if I was sober!" Drunk, they held each other up; sober, they got each other drunk. Bleary-eyed and unshaven, shirttails out and flies at two-o'clock-in-Petersville, half-pints bulging like tumors in their pockets, they lounged around town, uttering good-natured indelicacies in the presence of ladies in public places and performing their toilettes in broad daylight behind the skimpiest of bushes in the courthouse yard, corrupting the youth and bedeviling the established order, forgoing no opportunity to display the high standards of behavior that Needmore knew them for.

Needless to add, Harry thought them altogether charming.

"He's a-running the streets, Leona Pomeroy!" Granny was in the habit of declaring when, still chewing his last bite, Harry rose from the supper table already reaching for his coat. "He's a-running the streets with that old pool-hall trash, you mark my words if he ain't!"

"Oh *fiddle*, Mizriz Biddle!" Leona would demur.

"If this old hickish place had a little pep, they'd put in some good wholesome activities for their yewth! Why, up in Dayton . . ."

As Granny headed for the knitting bag, Harry headed for the door. Then off into the night for more boyish pursuits and diversions with his little chums.

Diversions, diversions. The preview for *Dads and Mothers* had arrived, and Harry and Mr. Ockerman watched it one Sunday afternoon while Harry popped popcorn. As a matter of fact, they watched it seven times. There was a good deal of footage of a pack of disconcertingly elderly-looking teenagers in bobby sox and saddle shoes, jitterbugging, cutting up in the halls of the local high school, riding in fox-tailed jalopies, drinking sodas in the Sweet Shoppe, and in general conducting themselves in the approved Andy Hardy fashion. The film would apparently focus on the tribulations of one irrepressible couple who suddenly find themselves, one moonlit night, on the far side of foreplay. The high point of the trailer was a brief shot of this enterprising pair smooching it up in an open roadster parked before a painted backdrop which featured a seriously out-of-round harvest moon. As the camera closed in on them, the precipitous lovers fell into a frenzied grope and slipped slowly out of view beneath the dashboard. A quick dissolve, then the screen filled with exhortations to "SEE the Horrors of Venereal Disease! SEE the Miracle of the Human Reproductive System! SEE the Birth of Swedish Triplets!" while a scratchy voice-over assured the audience that the Registered Nurse would be on hand at all showings, to minister to those who lacked a stomach for graphic representations of the fruits of illicit congress.

The footage was gray and grainy, the sound track hoarse with static, the clothing and hairstyles and automobiles suspiciously pre-war; and such acting as was displayed in the trailer would have suffered by comparison to a poke in the eye with a sharp stick. But Mr. O. was tickled pink. When these ignernt Amoses around here got wind of such goings-on as he and Harry had just witnessed, he predicted, the news would travel like grease through a goose.

And more diversions. On the very first Saturday night that Mr. Ockerman ran the trailer—as if in testimony to its aphrodisiacal power—Ramona Halfhill came up to Harry after the show and asked him flat-out to take her home. Harry accepted the proposal with almost unbecoming alacrity, even though she lived nine miles out in the country, and even though she was Ramona Halfhill; he needed the practice. Ramona waited in the car outside the theater while Harry got out the stepladder and changed the letters on the Now Playing sign to the next night's attraction—Gene Kelly in *High Steppin'*.

They parked in the lane back to the Halfhill farmhouse and immediately fell to, like a pair of famished cannibals striving to ingest each other by systolic action. During the progress of these exertions Harry once again laid hold, momentarily, of her left hickory nut—it hadn't prospered appreciably during their estrangement—but Ramona instantly jabbed him in the ribs so hard with a bony elbow that it felt as if she'd tried to drive a stake through his heart. While he was recovering, she observed (with greater accuracy, perhaps, than she knew) that "if I let you fool around with my b'zeer, Harry, you won't respect me as a girl." He tried blowing in her year again, but she said that little trick didn't move her one bit, and moreover that she had to get on in the house before her daddy came out and whipped her butt and—a vastly more compelling argument—Harry's, too. She did allow, however, before she left him, that his having a car had improved his personality immeasurably, and said she wouldn't mind going for another ride sometime. All thoroughly diverting.

Needmore had closed up shop by the time Harry got back to town after seeing Ramona home; it was even almost closing time at Craycraft's. Monk was there, sitting on the pop cooler sucking at a beer in a brown paper sack. Rod and Gene were lurching and reeling around the back table, essaying some dipsomaniacally obscure variety of pool. Claude, a sallow study in dyspepsia, sat scowling behind the counter with his triangle rack around his neck and Old Pismire, the poolroom's lank, moth-eaten yellow tomcat, in his lap. Old Pismire was as ill-tempered as Claude, and Claude was as yellow as Old Pismire—and those must have been the things they liked about each other, there being little else to admire about either of them.

"Hey, here's Step!" Monk called when Harry came in. "Take yer hat and jacket off, Step!" And then he said something for which Harry knew he would be forever in his debt. He said: "Step's a bad cat to clean after, boys! He's been out drivin' that nail again!"

"Step better watch his damn step," Claude grumped, "or Milton Halfhill will step on Step's little Buckeye ass." Claude didn't think any more of Harry than he'd thought of Harry's father. He banged the counter four or five times with the triangle, to get Rod's and Gene's attention. "Hurry it up back there, you two dingleberries!"

The Weave was just completing a very nifty two-rail bank into the corner pocket, which could only have been improved upon by his having shot the cue ball and sunk the five, instead of the other way around. As the cue ball dropped into the pocket, Rod seized his cue stick and delivered a vicious stab at the nine-ball—which article, so unceremoniously goosed, promptly leaped a good two feet off the table, cleared the opposite rail with inches to spare, bounced once on the concrete floor, and disappeared, with a juicy articulation attended by a little amber gusher, straight down the brass throat of an already overflowing spittoon. The doughty Rod, nothing daunted, wobbled resolutely and more or less directly over to the spittoon and thrust his hand into it to the wrist. When the hand emerged, it held, between thumb and forefinger, with pinky delicately extended, the wayward nine-ball, dripping a viscid elixir which had best remain otherwise undescribed.

"Oh, I gotchew now, you li'l booger," Rod remonstrated with the chagrined nine-ball, holding it up before him like Poor Yorick's skull. "Hey, Sharpie, we better get us a ssshtick an' kill this li'l booger!"

"What you two Sharpies better do," Claude corrected him, "is take that thing back to the tap and wash it off, and then haul your deadbeat asses elsewhere." He set Old Pismire, yawning and stretching, on the counter and began angrily wiping down the back bar with a rag.

"Aaw," Rod mourned, "we was playin' f' the champi'nship, man!"

"We was, Sharpie?" Gene marveled. "The champi'nshippa what?"

"Why," Rod said, tottering off toward the pissery, "the champi'nshippa two drunks, Sharpie!"

Gene, anxious to protect his turn at the half-pint of Sweet Lucy in Rod's hip pocket, tottered determinedly in pursuit, imploring permission to assist him. Monk and Harry got up to wait for them outside. The venerable Pismire favored them with a feral sneer as they approached. At the door Monk stopped and smacked his empty beer bottle smartly on the counter before Claude, and missed Old Pismire's tail by the scantest of margins. The indignant Pismire leaped to the floor and, at a safe distance, turned and arched his back and spat hatefully at Monk. Had Old Pismire and Monk each been, instead of themselves, their own respective ancestors at that moment, they would have answered, then and there, the age-old conundrum that small boys pose among themselves: Could a tiger whip a g'riller?

"How's about another cool one, dad?" Monk asked Claude, his manner blithe, airy.

"No more beer," Claude said. "That's it on the beer." Claude's matchstick danced a do-si-do along his lower lip, and his muddy, baleful eyes would have stared a tombstone—but not Monk McHorning—out of countenance.

"Right," Monk conceded breezily. "How's about a packa butts, then?"

The matchstick paused to take a bow in the corner of Claude's mouth, then resumed its mad cavortings. "Sure," Claude said. He tossed Monk a pack of Kools from the rack behind the cash register. "Watch them things, McHorning, they'll stunt your growth."

"Up yawz, dad," Monk said, pocketing the Kools. "Keep the change."

They went on out. Old Pismire, seeking diversions of his own, went out with them. Monk and Harry stopped at the curb and lit up. Inside, they could hear Claude ragging Rod and Gene to hurry the hell up and get the hell out, he wanted to close the hell up and get the hell home.

"Old Craycraft!" Monk mused, launching a smoke ring into the still, chill October night. "He's what you call one of these athletic supporters. He's gonna buy me a damn sportcoat."

"He is?" Harry said. "How come?"

"He never said—and I never ast him. I want me a one-button roll, dad, midnight blue."

While he was talking, Rod and Gene came tumbling and stumbling out the door. Claude locked up after them, turned off the neon sign, and started sweeping up.

"Claude's ridin' the rag tonight, ain't he," Monk suggested.

"Aw," Rod said, "ol' Claude's got liver troubles. You shouldn't run him down."

" 'At's a fack, Sharpie," Gene offered, "he's got the liver troubles. He's kind of a mustard color, ain't he."

"Aw yeah," Rod said solemnly. "You shou'n't run 'im down, McHorning. Me an' Sharpie's gonna have liver troubles one of these days. What if people runs us down?"

Laughing, Monk flipped his cigarette away and

turned to Harry. "You got yer short, dad? Let's take us a ride."

The two Sharpies were by this time under the streetlight in the middle of the street, each enthusiastically endeavoring to help the other don his coat—an interesting exercise in simultaneity even, when one stops to think about it, for two sober men. Rod and Gene were anything but sober, and for a moment there seemed a real possibility that they might break each other's arms. Monk and Harry got them untangled and maneuvered them into the back seat of the Hudson (Harry had some private misgivings about that arrangement, in view of Rod's reputation for spectacular projectile vomiting on intemperate social occasions such as this one), and Monk got in up front with Harry.

"Watch out for pecker tracks back there, men," Monk cautioned. "Step's been out drivin' that nail."

Harry was feeling, just then, exceptionally fine—and why not? He was a sport with a short, he drove that nail, his whereabouts were told in pecker tracks. And that was why, because he felt so exceptionally fine, out of sheer youthful exuberance, for the first time in his young life at the wheel, he scratched off. They burned rubber all the way up Courthouse Street; the rear end slipped and slewed behind them most refreshingly, and the agonized squealing of his tires afforded Harry enormous satisfaction.

"Where to?" he inquired coolly at the corner.

"Any goddamn where but here," Gene suggested, looking anxiously out the rear window. "You jus' run over Old Pismire, m'boy . . ."

TWELVE

Harry proceeded apace to the edge of town, where, to the huge relief of the front seat, the back seat unanimously requested permission to disembark and throw up and get it over with. That experience, however, together with the bracing effect of the chill of a late October midnight, sobered the two *bon vivants* intolerably, obliging them to propose that the entire company motor forthwith to Limestone and call upon a certain all-night bootlegger of Rod's acquaintance, who dealt in the fine wines to which he and The Weave were so profoundly accustomed, and that they pool their resources and invest in a couple more half-pints, so that Old Pismire could be memorialized with a proper wake. To sweeten the proposition, the back seat even offered the front seat the dregs of the last remaining half-pint, as a foretaste of the gustatory delights which all were shortly to enjoy. Harry hooked back his portion with great aplomb, but it went down hard, very hard. It tasted (the characterization was Monk's) like horse piss with the foam farted off, and for the second time that evening Harry's systolic apparatus was severely tested. But down it went, finally, and they were off, tooling down life's highway while the mortal remains of poor Old Pismire lay steaming on the silent stones of Courthouse Street, struck down by madcap Yewth, which, as even Pismire should have known, *will* be served.

They'd scarcely passed the Needmore city limits when Monk glanced behind him and said, almost (for Monk) affectionately, "Lookit them two fruitcakes, will ya." Harry found the dome-light switch and saw in the mirror that the two Sharpies, like a pair of unemployed bookends, had collapsed against each other in the back seat in a comatose condition. Heads flung back and mouths agape, shoulder-to-shoulder and cheek-to-cheek, they snoozed angelically.

"You know the thing about them two?" Monk said. "The thing about them two is, they could care less."

Harry didn't altogether understand the construction. "They . . . could?" he ventured.

"Damn straight," Monk asserted, not very helpfully. "Like you take old Inez, up at the Home." (Ah yes, the winsome Inez, who'd so discreetly advised Monk that he had one a-hangin'.) *Big* honker, Monk went on; had a head on her like a damn tarpeder. She wasn't none too bright, neither, he said, and she put out regular. Monk used to go into the broom closet with her; sometimes they'd get to thrashing and flailing around amongst the mops and mop buckets and brooms and brushes till you couldn't hear yourself think in there. So this one time they were going at it hot and heavy when old Inez gives him a big squeeze and whispers in his ear, "Oh, Monkey, you are so garjis!" Now Monk was only twelve years old at the time—Inez was sixteen or seventeen, bi-i-i-ig honker—but he had looked in the mirror a time or two, he knew what he was and what he wasn't, and garjis he was not. It was just a goofy idea in Inez's old tarpeder head, was all. Still, it was something of a satisfaction to discover that somebody considered him garjis, even if it was just big old dumb, lumpy Inez; so he puffed himself up and said (quite generously, he thought), "Why, Inez, dad, you're real . . . nice-looking yer ownself." But Inez wasn't going for it. "No I hain't," she'd sighed. (She came from down in the mountains, Inez did; a hillbilly from the word go.) "No I hain't, Monkey. I'm a ongodly ugly old fat idjit, is what I am. But I tell you one thing, Monkey"—and here she had taken him by the ears and held him out before her, so she could hoot directly in his face—"*at least I hain't got the damn big-head about it!*"

"Old Inez!" Monk said, chuckling fondly. "She could care less, dad. She mighta been bat-shit, but she learnt me a thing or three."

Harry waited, hoping for more detail about the cacophonous revels in the broom closet; but the particulars weren't forthcoming, and it didn't seem quite polite to request them. Instead, he asked Monk what ever become of old Inez.

"She took on the whole place this one time," he said. "When they caught her she had guys lined up outside that broom closet clear down the hall and around the corner and halfway down the stairs, and it sounded like a damn bowling alley in there! They shipped her out to the nuthouse. Said she was a skizzafrenny. But hey, she wasn't no damn skizzafrenny, dad, she was just a crazy nympho!"

Harry took the bull tentatively by the horns. "Speaking of big honkers," he said, "what about . . . ?"

"Hey, you know what Snakeshit says: 'When the weather's hot and sticky, that's no time for dunkin' dicky. When the frost is on the punkin, that's the time for dicky dunkin'!' Plenty of time, dad. Alla time in the world!"

It all reminded him, Monk went on, of the one about the fat lady and the plumber and the plumber's friend—which in turn reminded him of the one about the traveling salesman and the duck, which reminded him of the one about Oink Johnson, which reminded him of the one about the lumberjack, the knothole, and the woodpecker, which reminded him (the connection was obscure) of his recitation of the saga of King Arthur and the Knights of the Round Table, and by the time he was up to about the twenty-third verse of that epic ("Balls!" cried the Queen. / "If I had two, I'd be King!"), they'd topped the last hill and could see the lights of Limestone all strung out along the river far below them, glittering as if there really were something going on down there.

Rod and Gene suffered a rude awakening at the first traffic light in downtown Limestone, where a bit of jittery clutchwork (it was Harry's maiden traffic light) sent the Hudson hiccuping violently through the intersection and shook them from their slumber. They awoke much refreshed, and directed Harry through the deserted midnight streets to an establishment called Jack's Hack Shack, a corrugated tin hut situated in the alley behind the Kroger store, boasting one ancient cab which crouched, wheel-less and on blocks, in a shadowy niche of the narrow alley. But Jack's was a going concern withal. Its gnomish proprietor opened the door at Rod's first knock, and within moments Rod was back with four half-pints—for some unfathomable bootleggerly reason Jack dealt exclusively in half-pints—and they were on their way again. At the head of the alley the first of the new half-pints made the rounds, its contents already much reduced by the time it came to Harry. This time, in the interest of preventing another epiglottal seizure, he corked the bottle with his tongue before he tipped it up. At the last second, something in him wanted wine, and he let it have a little. For the moment it was mollified.

Thence to Stringtown, a foreshortened rat-tail of a community appended to the rump of Limestone. Harry cruised those vacant, dismal streets with his hand a-tremble on the wheel, imagining behind every drawn window shade either a bare-breasted *National Geographic* Hottentot or Lena Horne in a negligee. Twice more he fortified himself with a drop of Sweet Lucy, against an attack by renegade dusky damsels prowling the night in search of Caucasian-style romance. But once again, to the astonishment of the Sharpies, not the first smoked chicken rushed out to take advantage of them, and finally The Weave said What the hell, he'd sooner be home chasing his old lady around the bed, and they gave up and headed back to Needmore. For consolation, Harry offered himself another li'l drink —and thanked himself very much, and gratefully accepted it.

Harry had his hands full driving home. The highway, which, going the other direction, had been an ordinary rambling two-lane blacktop, persisted in separating and coming back together in the most unnerving manner, sometimes even weaving in and out of itself, twisting and writhing beneath the Hudson as if it wanted to throw it off and go somewhere else. Christ, thought Harry, clinging grimly to the wheel, first a traffic light, then a divided highway. But he rode it out for a few miles, until it settled down, and then he cleared his head with another sip or three of wine, and after that he really didn't remember much about the trip, which in any case was miraculously uneventful despite the fact that he drove the last eight miles entirely in the left lane. His companions took note of the eccentricity at the time, but they didn't mention it because, they cheerfully acknowledged later, they could care less —and anyhow, Monk said, at three-o'-goddamn-clock in the morning, what's to hit?

Harry did recall that when they got to town he congratulated himself with another li'l drink. Then somehow they were outside the car, and he was sitting unsteadily astride Monk's shoulders, and the words GENE KELLY swam before his eyes. There were more words, too, broken ones, and in his left hand he held, big as a six-gun, the letter "P."

"Now put that in front of the 'I,' " Monk coached hilariously from below, "and move the 'H' over between the . . ."

Back in the car, Monk rewarded Harry with one

more taste of wine. Then the wine was gone, and so was Harry.

He didn't rejoin the living until he woke up in his own bed on Sunday morning, much too sick to die. Later, when Leona and Granny got home from church, they said some old drunk must've crawled into the car and gone to sleep last night, because it smelt to high heaven, and they'd found an empty wine bottle under the seat.

Harry dragged himself from his bed of pain and drove uptown. The Now Playing sign on the New Artistic proclaimed:

GENE KELLY

IN

PIG SHIT

THIRTEEN

That the Bulldogs, in the season opener, handily dispatched the Vikings of Northern Kentucky School for the Deaf was no surprise to anyone, least of all to the Vikings themselves, who lost so often they must have wondered if the One Great Scorer was particularly ill-disposed toward the hard-of-hearing. But that the Dogs belabored the feckless Vikings to the tune of 92 to 17 on the victims' own home court, and that "this big old hammerheaded orphant boy" was personally responsible for exactly half the Bulldogs' total (which 46 points happened to be exactly twice the old Bulldog record of 23, set by Scudder Wallingford in 1941 against an earlier generation of those same much-abused Vikings)—these promising developments occasioned modest rejoicing in the ivied halls of Burdock County High and on the streets of Needmore. Heads were held a little higher, eyes shone a little brighter, hope flickered anew where hope had long since guttered and died; long-time Bulldog loyalists congratulated themselves for having known all along that Nobby Stickler was a damn fine little coach when he had the material. One or two bleeding hearts did murmur that it was perhaps not altogether charitable of the Mastermind of the Hardwood to leave his starting five in the lineup for the entire game, inflicting thereby a terrible humiliation on a brave band of handicapped children; but in the main these whinings were ignored in favor of Coach Stickler's own analysis, which held that the basketball bidness was no place for crybabies, and that if winning wasn't important, why did they keep score?

As for Harry, he could've cared less. What excited him about the game was that after it, on the bus, he got bare knee off Ramona Halfhill—an experience no less piquant for its brevity. He and Ramona were locking mandibles as usual at the time ("swappin' slobbers," Ramona called it, with relentless candor), and the instant his hand, seemingly of its own volition, dove beneath her voluminous satin-lined official Doggette twirly skirt she voiced a "MMMMMMM-PHTTT!" of muffled outrage and attempted with alarming purposefulness to bite off the tip of Harry's tongue. Harry withdrew both the offending extremities without delay, and when Ramona emphatically notified him, and everyone else on the bus, that if he tried that again he'd drawr back a bloody stump, he wasn't sure which appendage she meant to threaten. But all that notwithstanding, and notwithstanding that the knee was a poor, rawboned affair, as hard and unyielding as a joint of plumbing, the hand that had lain upon it for but the merest instant was warmed for hours afterwards by the memory.

Harry had been mostly lying low since the Limestone caper. His hangover alone had incapacitated him for days. (The first hangover is always the hardest, especially when Sweet Lucy has graced the revels.) Then there was the matter of Old Pismire, whose mortal person, under the pressing influence of the automobile traffic on Courthouse Street, was daily becoming less like a cat and more like the hairy leather shadow of a cat. Flat as a cow-pie and stiff as a boiled shirt, caught now forever in his customary cringing attitude, he silently reproached Harry from his gutter every time his assassin crossed the street.

Worse, when Harry, after discreetly absenting himself from the premises for almost a week, ventured at last to drop by the Billiards one evening, Claude Craycraft motioned him over the minute he stepped inside and advised him that he was, in

Claude's estimation, a half-assed Buckeye with skidmarks in his skivvies, and he just better watch his damn step, was all. Harry slunk away without contesting Claude's assessment of his character and situation, and steered clear of the poolroom for a few more days.

In their home opener, the Bulldogs treed the Toole County Tomcats before a howling audience of Bulldog Boosters. This time the margin was a less imposing, and more realistic, 47 to 36, but Monk scored 32 himself, and the two victories in a row constituted Burdock County's longest winning streak in years. BEWARE OF THE DOGS! screamed the headline in that week's edition of the *Bulldog's Bark,* above Harry's account of the unique brace of stirring triumphs.

On a Saturday afternoon a week before Dr. Rexroat's engagement, Harry staggered about the streets of Needmore for several hours under an enormous bundle of the doctor's handbills, slipping them under windshield wipers, tacking them to telephone poles—and at last dumping a substantial number down an outdoor toilet at the municipal baseball diamond. The advertising campaign received a nice boost the following morning, too, when a number of local ministers took to their pulpits to denounce sectual hygiene as the Devil's work. Granny arrived home from church in high dudgeon, fuming that in her opinion this old hygiene was just so much nasty dirty filth, and how she just simply did not *see*, Leona Pomeroy, how you could let a boy of your'n . . . But Leona countered with a stout Oh-fiddle-Mizriz-Biddle, and enjoined her not to be such a stick-in-the-mud, and Harry saw that Progress had already won the day.

The great day dawned gray and chilly, but that didn't stop Harry from washing the Hudson all morning long. The frost was definitely on the punkin, he cheerfully observed, noting how his hands turned blue with cold each time he rinsed his chamois.

At noon he slipped quietly into the house—Leona was still asleep—and fixed himself a couple of fried baloney sandwiches. He ate hurriedly, because Mr. Ockerman had asked him to come in early that afternoon, to work up a triple supply of popcorn and help make the theater ready for the evening's festivities. Granny, charging *in situ* in her rocker like Ben Hur in his chariot, entertained him during his meal with a grim litany of mutterings regarding filth and dirt and nastiness, while Pittybiddle chimed in with a brisk roll call of all those self-respecting McAtees and Pomeroys who, presumably, would have had no dealings whatsoever with such unsanitary affairs.

By midafternoon Harry had popped and bagged a very Matterhorn of popcorn, and had just finished wiping down the popper, when his employer came steaming in, excited as a bride.

"Harry," Mr. Ockerman exulted, "these prunish

old women and preachers and whatnot has just about run me ragged the last few days! A-calling me up at all hours of the day and night, as late as eight and nine o'clock of an evening, telling me how if my daddy was alive he would turn over in his grave and whatnot! What in the h-e-double-l are you all talking about? I says. You all are messing with the legitiment theater, I says, don't you know that? Culturewise, Harry, certain persons don't know culture from a hole in the ground. But the show must go on, Harry, am I right or wrong? The show *must* go on!" Mr. Ockerman shouldered his broom and marched off down the aisle, Harry dragging the trash can in his wake. "And by dammit she *will* go on, won't she, Harry, if you and me has anything to say about it! Shoot, I wisht Dad *was* alive! He'd be thrilled to death!"

Warmed to the task by these and countless similar effusions, the two showmen pitched in and made the dust fly for an hour or so. Then they fired the furnace, put new bulbs in the footlights, drew the stage curtain, and even managed a halfhearted pass at swabbing out the rank, dank toilet under the stairs. As they labored (except during that final chore, which rendered him temporarily breathless), Mr. O. whistled "There's No Business Like Show Business" whenever he wasn't talking.

FOURTEEN

Harry parked the Hudson in the alley behind Blankenship's that evening, instead of in front of the theater, so that—having already enjoyed *Dads and Mothers* in the exhilarating company of the belles of Burdock County—he could steal away quietly during the men's showing, swing by and pick up Oodles, drive her the two miles out of town to the Highway Department's heavy-equipment lot, park behind the rock crusher, give Oodles the thrill of an old maid's lifetime and dispose of their mutual virtue in, so to speak, a single stroke, and be back at the theater in, oh, about half an hour. Allowing for foreplay, forty-five minutes at the outside.

Discretion was of some real consequence to Harry in this operation. He and Ramona had been working out together pretty regularly of late, and though there was undeniably a certain clinical, detached quality about their ardor—almost as if they were engaged with two entirely different people, it usually seemed to Harry—he'd come to look forward to these overpopulated trysts, and he would've hated like anything to have to give them up. The palm of his right hand still burned when it recalled that knee. He was getting up his nerve to try that again sometime soon.

But the Halfhills wouldn't be in town tonight, there being no cowboy movie to bring the little Halfhills to. It had briefly occurred to Harry a few days earlier that he just might be forced to tool out to see Ramona after the second show, and try for a double-header. That would have been a real accomplishment, to cross two virgins—no, three—off the list in one night. But when he'd hinted to Ramona that he'd considered gracing her with his company after work on Saturday evening, she disabused him of the fancy in a hurry, declaring that for his information he neen't to suppose he could come bothering around her after he'd went and got sexed up at some old dirty show. She had enough trouble with him as it was, she said. And anyhow, she said, their neighbor had a new teevy, and she intended to go over there and watch Gorgeous George on it.

It was barely six-thirty when Harry let himself into the theater, but Mr. Ockerman was already there. He expected Dr. Rexroat at any minute. In a fit of opening-night jitters, he was pacing the floor and wringing his hands as if Mrs. O. were about to give birth to the Swedish triplets herself, and Dr. Rexroat and his nurse were coming to deliver them.

Harry had just hung up his coat when they heard someone rattling the outer lobby doors. Mr. Ockerman attended and came back beaming, accompanied by a wizened old sport wearing an outsized, off-white Palm Beach suit in an advanced state of deterioration, a black beret plopped onto the side of his head like an enormous jelly pancake, and—it's mid-November, bear in mind, liable to snow any day now—muddy white perforated wingtip oxfords; also a hulking, horse-faced woman with oakum hair, a lowering brow, and a cigarette jammed in the corner of her mouth like a vent pipe for the fumes of something smoldering inside her. The woman wore dirty pink pedal pushers, a magenta sweater, baby-blue anklets, red plastic spike-heeled pumps, a gory smear of lipstick, and a yellow fur jacket that looked as if it might have been constructed of the pelts of about twenty-two Old Pismires. She was a Technicolor extravaganza, her companion a Selected Short Subject in black-and-white; the only color he showed was in his ascot, which was a rusty montage of chili spots, and in his eyes, which were as bloodshot as open wounds.

These imposing personages, Mr. Ockerman announced, were the illustrious Philander C. Rexroat, Sex.D., and his able assistant, Miss Ratliff, R.N.

The doctor's hand, when he offered it to Harry, was as dry and frail as a dead leaf, and trembled violently. "Thuch a pleathure to thee you," he rumbled in a voice so deep and resonant, despite the muted tone and the lisp, that Harry could hardly believe it had issued from this desiccated little leftover of a man. He hadn't a tooth in his head, so far as Harry could determine, and the atmosphere about him was shot with the juniper scent of gin.

"We, ah, thtopped by your local café for a bite of thupper," the doctor was explaining now to Mr. Ockerman. A snippet of gray mustache clung like lint to his upper lip. Like his lady, the doctor was a smoker; he affected a long black cigarette holder, which he managed with some difficulty to hold clamped between his gums. It was as if a Hollywood Svengali had tried to turn Gabby Hayes into a matinee idol. "And, ah," he continued, "thomeone there directed uth . . ."

"Jayzus P. Christ!" Miss Ratliff bawled, in tones not notably prayerful. She was stalking up and down the lobby, peering into the darkened theater and laying down smoke as if she were overdue for a tune-up. "The son of a bitch is another goddamn fire-trap! Wouldja look at this goddamn son of a bitch? I bet the son of a bitch ain't even got a goddamn johnny, and me about to bust!"

Who Registered that Nurse? Harry wondered as Mr. Ockerman ceremoniously bowed her into the facility beneath the stairs. Harry and Mr. O. went outside with the doctor to help him unload his car. On the way out, Harry noted by the lobby lights that the distinguished doctor could have used a haircut and a shave, and that the back of his neck was no cleaner than it ought to be. His overall complexion was basic tattletale gray.

Dr. Rexroat's car was as much too small as his suit was too large. A refrigerator-sized midget auto manufactured by a company—the Crosley Corporation—whose principal product was in fact refrigerators, it cowered by the curb as though it feared some real car might come along and cast aspersions on it. On the little car's door was a crudely executed heraldic shield, encircled by the words "Rexroat Productions," which bore the head of a unicorn (it could also have been a one-horned German shepherd) and the motto *Omnia vincit amor.* Dr. Rexroat opened the door and crept inside like a white mouse entering its hole, and began pawing with mouse-like frenzy through the disordered heap of boxes and cartons, film canisters, odd pieces of small luggage, empty gin bottles, wads of clothing, loose handbills, mateless shoes, and nondescript flotsam and jetsam that filled the tiny back seat to the gunnels. Miss Ratliff had pressed the entire front compartment into service as her personal ashtray; it was ankle-deep in lipstick-smeared cigarette butts. The doctor's meager hindquarters held the door ajar; the seat of his trousers, Harry saw, was polished to an iridescent threadbare sheen. He wore no socks, and there was a dime-sized hole in each of his shoe soles, a pair of mournful eyes mutely testifying to hard times in the sex-hygiene business.

Dr. Rexroat extracted from the Crosley, without seeming to have reduced or even disturbed its total contents in the slightest, a carton of books, three large film canisters, a rumpled, once-white hospital smock and an equally ill-used starched white cotton nurse's cap emblazoned with a misshapen Red Cross rendered in fingernail polish, a shoe box full of mostly empty medicine bottles of various description and prescription, a tin cashbox, and two framed certificates with impressive gold-foil seals, the one attesting that Philander C. Rexroat had completed with distinction the course of study leading to the Doctorate in Sexual Behavior at the Instituto de la Psicología Humana of Nuevo Laredo, Mexico; the other that Wanda Pearl Ratliff was a graduate of the Lester Mullins Junior High School of Ardmore, Oklahoma, class of 1931. Also a fresh pint of gin, which the doctor surreptitiously slipped into his inside coat pocket.

They lugged this store into the lobby just as Miss Ratliff emerged from the chamber beneath the stairs, delicately holding her nose and crying, "Pee-yew! Mister, your goddamn johnny stinks like a son of a bitch!" Mr. Ockerman looked somewhat injured by this intelligence, but her report did not in the least deter Dr. Rexroat, who said, discreetly patting his inner pocket, he believed he'd, ah, jutht thtep into the, heh-heh, johnny himthelf and, ah, frethen up a bit.

"Hey, Phil," Miss Ratliff cautioned gruffly as he scuttled past her, "don't you drink tee many martoonies, now. And put them goddamn choppers in, you look like billy-hell."

In the doctor's absence Harry set up, at Miss Ratliff's direction, a card table and a folding chair near the exit, and Mr. Ockerman hung the two certificates on the wall behind the table—"back in the corner, honey, where there ain't so much light, if you follow me." The doctor rejoined them shortly, wiping his mouth with the back of his hand. He had assumed a pair of impenetrably dark Hollywood-style sunglasses, in which he looked more than ever like one of the three blind mice. His choppers were in place, and he demonstrated them by way of a death's-head smile. His hand seemed a bit steadier, but, as if in counterpoise, he was now a little wobblier on his pins. In passing, he inspected Miss Ratliff's first-aid setup and ventured the opinion that her medicine bottles might look just a smidge more convincing if they, ah, had a little something in them, and that "this charming young man"—that was Harry—be dispatched up the street to the filling station to fetch a root beer. The doctor's lisp was gone, owing no doubt to the installation of the choppers.

When Harry got back with the root beer, the doctor was down front setting up his speaker, and Mr. Ockerman was in the ticket booth counting change for his and Harry's cashboxes. Miss Ratliff had assumed her nursely habiliments, having set the crushed Red Cross tiara at a jaunty angle on her scraggy head and put on, the lobby being, she'd observed, one goddamn cold drafty son of a bitch, the exhausted hospital smock over her fur coat. She was a burly, barrel-chested old skagmaw, Nurse Ratliff was, with the bosom of a beer-truck driver; and crouching there over her table decanting root beer into her mysterious vials, with tufts of Old Pismire peeping through her gown at every rent and seam, she might have been Dr. Jekyll's lab assistant, or the Florence Nightingale of the Neanderthals.

Her cashbox was on the table, along with a stack of cheaply bound paperback volumes entitled *The Illustrated Scientific Encyclopedia of Human Anat-*

omy and Sexuality, by Philander C. Rexroat, Sex.D. A small sign indicated that these were available to the public for two dollars and fifty cents a copy. Harry reached for one, to examine it more closely, but when Nurse Ratliff saw him she set down her root beer and sharply smacked his hand.

"Hands off, sonny boy," she said. "Them's for sale. This ain't the goddamn lending liberry."

Down front, Dr. Rexroat rumbled "Tethting . . . tethting . . ." into a nonexistent microphone. He was projecting magnificently, but the choppers had evidently found their peripatetic way back into his pocket.

"That old son of a bitch," Nurse Ratliff said, more fondly than reproachfully. "I have to watch him like a goddamn hawk. You keep an eye on him too, hon," she cautioned Harry. "He can't hardly tell the boys from the girls anymore."

Some sexologist, Harry thought.

A few minutes before time to throw open the doors to the ladies, Harry helped Mr. Ockerman carry the film canisters up to the projection booth. As they made their way up through the balcony, they could look almost straight down on Dr. Rexroat, who'd chosen that unpropitious moment to step just into the wings to fortify himself with another little nip. "You know, Harry," Mr. Ockerman whispered as, far below them, a miniature Dr. Rexroat knocked back his little nip so hard that his beret fell off, "if

I didn't know he was the Cecil B. De Mille of Sex Hygiene Entertainment, I never would've believed it."

Harry wasn't worried. As a matter of fact, his excitement had only just begun to quicken at the imminent prospect of watching sectual movies with a mob of similarly excited ladies. Who knew what might happen? Maybe they'd run riot, pounce upon him in an Amazonian seizure of carnality and rip his clothing from his back and use him horribly. And even if things didn't come to such a satisfactory pass, just the very fact that they'd all be sharing this stimulating experience implied a certain intimacy, a common stirring of the blood. Harry had already practiced—at home, before the bathroom mirror—the insinuating look, the sly wink with which he intended to let Oodles know after the show that he was privy to the innermost workings of her unconscious mind (in Oodles's case, the only mind she had), and that he had a few ideas of his own, when it came to that. He'd learn her a thing or three!

So resolved, Harry set his film canister next to the projector, and was turning to go downstairs to face the virgin onslaught, when Mr. Ockerman said he guessed they wouldn't be needing him any longer, and maybe he'd better run along.

Run along?

Why yes, said Mr. Ockerman. Just till time for the men's showing, he meant. Because the first show being ladies-only, y'know, it wouldn't look good if . . . He had just took it for granite that Harry understood he couldn't . . . So Nadine would be there in a minute to catch the ladies' tickets, and handle the popcorn and whatnot, and if Harry would just try and be back by about twenty to nine . . .

In a matter of minutes Harry had his coat on and was standing on the sidewalk outside the exit door, holding it open for Oodles, who was going in as he was going out. All giggle and jiggle and titter and twitter and waddle and twaddle, training a lilac spume of Eau de Tutti-frutti so dense Harry could fairly feel the droplets condensing on his eyeballs, she swept past him as if he weren't there—". . . so I go Aw Poppy, what's the *diff* if little Harry sees it with the gals? And Poppy goes he just ditn't like the *ide*a, beings as he's so *young* and whatnot, and Mommy . . ."—leaving Harry planted out there on the sidewalk like something she'd stepped in, automatically winking his sly, insinuating wink again and again at the door as it swung shut behind her.

He stepped over to the curb and administered a rattling good kick to the right front fender of Dr. Rexroat's Crosley, and then he skulked off up the street, glowering into the swiftly gathering dusk. Suddenly his own elongated shadow fell before him on the sidewalk. Oodles had turned on the outside lights; the World's Greatest Sex Hygiene Attraction was open for business.

FIFTEEN

The rest of Needmore, on the other hand, was to all intents and purposes out of business.

Harry hadn't been at liberty to go uptown at that hour on Saturday evening in a long while, but not many months ago the streets and stores would have been jammed with shoppers and socializers, and Courthouse Street teeming knee-deep in small fry till past seven-thirty, when they began convening in the New Artistic to cheer on the Westward Movement. Tonight, but for the little congregation in front of Hunsicker's Hardware, staring fixedly at an ant-like swarm of midget wrestlers on the Sylvania's tiny screen, the street was deserted and forlorn, desolate. Places that used to stay open on Saturday night till nine-thirty or ten—the barbershop, Blankenship's, the Self-Serve, even Hunsicker's itself—were already dark. The White Manor Cafe—the White Manure—was open, and of course the Billiards, and the lights were still on in Conklin's. Harry considered killing a few minutes in the drugstore brushing up on the progressive doings in *God's Little Acre*, but even as he was deciding, the window suddenly went dark, and Marvin Conklin came out with his topcoat on, and locked the door behind him. If prosperity is just around the corner, Harry mused, it must be on its way out of town.

They didn't care for loafers at the White Manure, and Harry had the feeling that Craycraft's might break his concentration. For a few minutes he joined the dumb-struck wrestling enthusiasts outside the hardware store, but soon his feet got cold and the rest of him got restless. Finally he remarked aloud that the midget whose leg was being broken wasn't suffering very persuasively, and the big, sour-faced bozo nearest him growled that if he didn't like the show he might consider shoving it up his old rusty-dusty. Harry reluctantly took cover in the Billiards.

Such trade as was being turned that night in Needmore was being turned in Craycraft's. There were lively games of check-pool on both front tables, nine-ball on the back table, and a four-handed euchre game in the booth by the pinball machine. Beer drinkers lined the bar two-deep. A party record called "She's Got Freckles on Her But . . . She Is Nice" was playing over and over on the jukebox, courtesy of a woebegone old soak named Juicy Morgan, who'd put in four quarters and punched it twenty-four times—because, he kept explaining, it reminded him of an old girlfriend of his.

Harry took a stick in the nine-ball game with Swifty and Foots and Norval Stroud, but he couldn't keep his mind on it. Although Monk hadn't come in yet, he was present in spirit as the principal subject of a great deal of optimistic speculation about the Bulldogs' prospects in next Tuesday's contest with the Lions of Limestone High, which would be the first serious test of the revivified Canines' mettle. Claude had taken Monk shopping in Lexington that afternoon—"because Stickler's too damn chinchy to buy him a decent outfit"—and he had had a good long talk with him, and he was satisfied the boy had come to play, and would win some damn ball games for a change. Even so, Claude said, as a public service he was willing to cover any little sums his regular patrons might care to set at hazard on the Limestone game. (Naturally, he added in passing, he'd have to have five or six points . . .) But even Monk's most devoted admirers, still adjusting to their new reality, still tremulously discovering what it was to have not just hopes but expectations for their Dogs, were wary. They all made much of their readiness to bet a million dollars, or ten million, or a suit of clothes, or their left testicle or some other expendable anatomical feature, but nobody wanted to play with cash money.

At the nine-ball table Swifty and Foots and Norval kept pestering Harry for inside info about the movie —particularly about the Registered Nurse. Was she a looker? Did she put out? Big knockers, huh? Big knockers? Hell yeah, Harry assured them, she's got 'em out to here! (*Down* to here, he corrected himself privately, feeling surly.) He hung up his stick and retreated to the pop cooler, above the fray.

The Lower Element was exceptionally well represented, even for Craycraft's on a Saturday night. In addition to the usual solid citizens, there was a large complement of rough customers and hard cases tonight, real lowlife: deadbeats and tosspots and hometown hardasses, smelly and sullen. Also stirred into this unsavory stew were five or six irascible-looking eyesores from Shankton, a tough little rock-quarry town in the southern corner of the county, and another five or six from Latchburg, an equally mean river town in Crawley County, just across the Burdock line. Stout fellows all, the salt of the earth, rounders and rips and ruffians, the ruck and the rabble, the riff and the raff.

Yet they were patrons of the arts, these waggish lads, proponents of the free and open exchange of ideas, lighters of candles rather than cursers of darkness, implacable foes of ignorance and vernal disease, apostles of sectual hygiene—for almost to a man they were getting drunk or staying drunk or getting drunker in preparation for *Dads and Mothers.*

A number of advanced thinkers were presently engaged in the idle persecution of a notoriously chuckle-headed party named Merle Toadvine, who was allowing his work-weary old wife to go to the movie. Hey, Toad's slick as a minner's peter! they chortled, nudging one another. Sent his old woman to the dirty pitcher show, so she'll . . . ! Toad acknowledged the tribute to his genius with a modest blush and a sheepish grin. Why hail, he said, tipping back his hat to scratch his head, she's done lost her nature, I can't hardly get her started, any more. Harry felt a little sheepish himself.

At the bar, Pillbox Foxx, a solitary holdout, was declaring them all out of their disgusting little minds for paying to watch a farrowing, he didn't care if she was a Swede or a damn Duroc. As a member of the vet'inary p'fession, he said, he could tell you stories that'd . . .

Harry was uneasy. The Billiards' crowd reminded him of something in himself just now, something unseemly and depressing. He remembered Dr. Pinckton's patient, the unfortunate Mr. Brown, who once stood, as Harry stood now, at the brink of the abyss. Brown had taken the plunge, and had emerged a ruined and broken man.

Also, Dr. Pinckton's fellow sexologist, Dr. Rexroat, and Nurse Ratliff hadn't exactly inspired Harry's confidence in them as Agents of Desire. He was doubtful that any love potion brewed by those two metaphysicians would have the desired effect, even on Oodles. Nor was he entirely comfortable about sharing a strategy of seduction with Merle Toadvine.

And even if the strategy worked, it didn't seem quite . . . sporting, somehow. There he was, slim and youthful and loaded with charm, a sport with a short; there was Oodles, putting on the pounds, none too bright, not getting any younger. He oughtn't to need outside help; the advantage was all on his side as it was. Then too, he hated having to betray Mr. Ockerman, he really did, whenever he happened to think about it. And lately, when he tried to call back to his mind the image of Oodles at her bedroom window, what came inexorably to the fore instead was Ramona Halfhill's old knobby knee. Maybe—another depressing thought—he was in love or something.

Monk showed up a few minutes before eight, resplendent. Claude had outfitted him in the sportcoat of his choice—one-button roll, dad, midnight blue—and had thrown in a pair of pale gray sharkskin slacks, baggy at the knee but pressed to a sharp twelve inches at the cuff, and pointy-toed blue suedes with metal heeltaps. The coat was cut very long, in the style known as the drape shape, with masses of blocky padding in the shoulders and wide lapels tapering down to a single button well below the waistline. The principal effect was to render him more ape-like than ever. But its like had not been seen before in Needmore, and Monk entered the poolroom to a rousing chorus of compliments on his good taste.

"That Craycraft's a live one, dad," he said to Harry, lovingly fingering a lapel. "This nigger-rigging run him over sixty bills, and he didn't blink an eye."

That's a live one? thought Harry, glancing at Claude. He looks like a stuffed alligator—and he *never* blinks an eye.

"I seen yer short up the alley." Monk was selecting a cue stick from the rack on the wall. He treated Harry to a leer and a grin. "Gonna drive that nail tonight, are yez?"

"Welp, uh—" Harry began, but Monk had already turned away to plunge into the check game. Well what? Harry wondered bleakly. Mired to his eyeballs in iniquity, deviousness, and coarseness, he longed, suddenly and overwhelmingly, for honor, for something he could claim to have won on his own merits.

At the check table, Monk already had a run going. He was playing well and flamboyantly, a drape-shape show pony clack-clack-clacking his heeltaps round and round the table, dropping everything he aimed at. All eyes were upon him. It was the right time to steal quietly away, and Harry seized it gratefully. As he'd feared, Craycraft's wasn't doing his concentration any good.

Outside, the little gathering of the faithful in front of Hunsicker's was just dispersing, Herb Shriner having only moments ago slipped suddenly into whatever dimension obtained on the far side of the screen. "Hit wadn't over yit!" wailed the wrestling aficionado of Harry's late encounter, as Harry passed him on the sidewalk. "If I had me one," he vowed, "I'd shet hit off when I dern pleased!" Mr. Hunsicker, snugly settled before his own Sylvania at home in his cozy living room, must have been smiling to himself right then.

Harry killed a few minutes sitting in the car behind Blankenship's, smoking and thinking, sorting things out. It was beginning to occur to him that he'd counted on *Dads and Mothers* to do as much for his own capabilities as it was supposed to do for Oodles's appetite. Now she'd be all hot and bothered, and he was . . . not. His concentration, he told himself grimly, was shot to hell. He clenched his teeth and held his breath and tried to bear down, but it didn't work; he was unmanned. With his deliverance from childhood almost at hand, he found himself beset with apprehensions and misgivings. Like a little boy about to suffer his first haircut, at the last minute he couldn't quite convince himself that he was ready for the operation.

By a little before eight-thirty he was back at the theater, wondering whether he dared try to slip in early. There weren't any more parked cars on the street than had been there before; evidently, throngs of sex-crazed women had stayed home with their husbands, where the action was. Harry stealthily unlocked the exit door and opened it a crack, then eased inside and closed it softly after him.

The houselights were up, and Dr Rexroat was on stage, addressing the dozen ladies who clung together in little coveys here and there about the auditorium. Oodles sat on a tall stool by the popcorn table, her immense, deeply upholstered back near enough that Harry could've reached out and snapped the strap of her b'zeer. She was wearing a shaggy, sleeveless, white angora sweater; from the rear she looked like a polar bear with its forelegs shaved. Nurse Ratliff had taken refuge in the johnny once again; cigarette smoke was rolling out from under the door, and Harry could hear her muttering within on the subject of why the goddamn son of a bitch wouldn't flush. The floor around Oodles's stool was littered with wadded popcorn bags. Even as Harry watched, she crushed another empty in her right hand and dropped it to the floor while reaching for a fresh bag with her left. But she was listening raptly to the doctor—a good sign, Harry figured; she was probably hot and bothered.

"Ah, Woe-man!" Dr. Rexroat was waxing boozily

enthusiastic. "She is a veritable . . . a veritable vessel of delights . . . a treasure trove of pleasures . . . a fount of fascinations . . . a precious chalice . . ." The Man of Science was listing dangerously to the left, but his voice rolled on, a treacle avalanche. "She is the repository of ecstasies both given and received . . . the, ah, seedbed of all our hopes and dreams . . ."

In the auditorium the several attendant vessels, troves, founts, chalices, repositories, and seedbeds accepted these encomia with a smattering of nervous giggles, perhaps uncertain whether they'd been complimented or insulted. Oodles, a jumbo marshmallow bonbon, wriggled prettily atop her stool. Harry felt his concentration surging back.

Dr. Rexroat lurched on, entreating mesdames not to submit the, ah, crucible of civilization to the blind and, ah, savage forces of ignorance and superstition. As an eminent scientist himself, he assured them that only in modern medical science could the gentler sex find sanctuary from ill usage, disease, pestilence, and female troubles, and, as a noted theologian and man of the cloth, he had it on the very highest authority that in the eyes of God, the only sin is ignorance. And as it happened, among the many educational activities of the Philander C. Rexroat Scientific Foundation for Sexual Enlightenment was the publication of educational material of special interest to the thinking woe-man—most recently, a limited edition of *The Illustrated Scientific Ency-*

clopedia of Human Anatomy and Sexuality, penned by himself, Philander C. Rexroat, Sex.D., with sections on the Twenty-seven Positions of Love, the Ten-Point Guide to Complete Satisfaction, the Eleven Most Sensitive Areas of the Female Body (Eleven? Harry marveled, staring hard at Oodles's ample back and counting furiously), the Four Ways to Regain Lost Sexual Power ("Show this one to the hubby, girls!"), plus Dr. Rexroat's Own Sure Cure for Menopause, and a special section on the Intimate Sex Secrets of the Hollywood Stars, the whole lavishly illustrated with full-color plates, all this and much, much more for a trifling two dollars and fifty cents the autographed copy, marked down from nine ninety-eight, not for sale at any bookstore or newsstand, but available while they last, strictly on a first-come-first-served basis, at the table in the lobby operated by his, ah, lovely assistant . . .

A great whoosh rumbled through the johnny's innards, the lovely assistant eminent scientist having just mastered the flush mechanism. Harry slipped back out just as she emerged, goddamning and son-of-a-bitching vigorously under her breath.

Outside, half a dozen men and boys were assembled now near the lobby doors, stamping their feet and blowing on their hands—and passing around a half-pint of whiskey, as a precaution against chills.

"Hey, Eastep," called one of the boys when he saw Harry, "y'all got any ginch in there?"

"Yeah," Harry said. "Your mother, Jimson"—a mean little jape which had the added virtue of being true. Jimson—who was smaller than Harry—beat a hasty retreat, amid further jollities from his fellows. Harry, transiently cocky, ambled over to the curb, propped a foot on the Crosley's fender, and lit a smoke. If a regular-sized woman had eleven Most Sensitive Areas, how many could he reasonably expect to find on Oodles? All concentration now, he was up to sixteen and still working on it when the door opened and the ladies began trickling out.

Among the first was the widow Jimson, whose wily offspring attempted to hide from her by hunkering down among his companions—who promptly betrayed him, whooping "Here's Elmo, Miz Jimson!" and even "We'll hold him for you, Miz Jimson!" She dove into the crowd and captured the miscreant and marched him off up the street, holding him so firmly at the armpit that his feet barely touched the sidewalk. "You go set in the car, Elmo Jimson!" she commanded. "You're too little to see that old pewkish pitcher!"

A trio of farmwives came out looking so profoundly scandalized that they intimidated the men into temporary decorum. Then came a hoary pair of sandwich-bar drabs from Latchburg, ardently thumbing through *The Illustrated Scientific Encyclopedia of Human Anatomy and Sexuality*. "Hey, Rosemary," called one of the male Latchburg contingent, to a rising chorus of guffaws from his ranks, "you ain't studyin' up on new ways to ficky-ficky, are you there?" Rosemary stuck out an antique purple tongue at him and snapped, over her shoulder, "I shore never learnt nothing new from you, Darnell!" She turned back to her friend and the book. "I don't see no full-color plates, Marcella, nor no Hollywood stars nor nothing," she was saying as they passed Harry, "but law, didn't that doctor have the prettiest way of talkin'!"

The crowd of men and boys had swollen by now to fifteen or so. A Dutch concert of wolf whistles greeted several of the ladies, and the half-pints were circulating as fast as dirty jokes. Harry made his way over to the exit door just as it opened to emit one last lady, Mrs. Merle Toadvine, bent and furtive, with frightened eyes and her face buried in the collar of her worn coat—and a copy of the *Encyclopedia* clutched to her frail bosom.

"You take you some Hadacol for that goiter, hon!" Nurse Ratliff called from her station in the lobby. "It done my sister's gallbladder a world of good!" Harry caught the door and stepped inside. "That pore thing!" Nurse Ratliff lamented, smoothing Mrs. Toadvine's two crumpled dollar bills on the tabletop. "A person sees the pitifulest things in the nursing trade. If I didn't love *people* so goddamn much, I would've went into some other line of work."

Dr. Rexroat was also at the nurse's table, autographing a copy of the *Encyclopedia* for Oodles, who stood hard by, talking like a teapot as she counted out ten quarters under Nurse Ratliff's watchful eye. "I mean I don't give a poot for this old s-e-x myself, beings as I am a single gal, you know, but I have never knew anybody which had wrote a *book* before, so I just go Boinnng, Deenie, you have *got* to have you one!" Oodles's lacquered platinum pompadour—another Shirley Worthington production—rode her brow like the hood ornament of a dirigible. Harry couldn't help noticing that there were no quarters in the popcorn cashbox.

"Here you are, m'dear," said the doctor, signing the flyleaf with a swordsman's flourish. "And may I say I do hope you get a great deal of, ah, satisfaction out of it, as the years go by." The rest of the doctor was now as mellow as his voice; he was blissfully ginned behind his shades.

"Why, *thank*yew, kind sir!" Oodles burbled, essaying an elephantine curtsy. "You know it is just like I was telling Mommy, I go Oh Mommy ain't I lucky, I am rilly gunna meet a real Hollywood movie p'ducer, because Mommy always says *I* ought to go into the movies, but I always go Oh Mommy ha-ha-ha it woultn't be the thing for me, I mean I honestly don't believe I would *make* it, for you know there are so *many* beautiful gals out there, no indeedy-weedy, little Deenie will just set right here in Needmore and be a big feesh in a small pond, ha-ha-ha."

While she was talking, the doctor had put forth his hand—now steady as a surgeon's—and was absently stroking her shaggy shoulder, petting the polar bear, murmuring "Mmm-hmmm, mmm-hmmm."

"Old man!" Nurse Ratliff snorted behind him, rising noisily to her feet. "You better go pee, and fergit it!"

Dr. Rexroat, his hand already going for his gin pocket, hastily excused himself and fled into the johnny. Nurse Ratliff muttered something about needing a breath of fresh air and stalked in after him—an indifferent place to seek fresh air, it seemed to Harry—and slammed the door. On the far side of the door she goddamned and son-of-a-bitched the doctor to a turn.

Mr. Ockerman would be up in the projection booth for a few more minutes yet, rewinding the film for the second show before he came down to sell tickets. Oodles, still a-blither, was putting on her coat. She didn't seem to have noticed that Harry was there at all, even though he was now her only audience, and had sidled so close to her that she was talking directly into his face.

". . . and Mommy goes Now, Deenie, if you *do* go into the movies, I certainly hope you won't be one of those old *sex*pots, missy . . ."

"Um, say there, Oodles, howdja like t'gofer a li'l, heh-heh, ride with me in my, heh-heh, short . . ."

". . . so I go Oh Mommy you *know* that is not my style, I mean I am more of the Greer Garson— What,

Harry? I do wish you woultn't always mumble your words, Harry. Have you got an insecuriority complex, or what? And speaking of you, Harry, how *is* your dear friend Mr. McHorning these days? He is so *aw*ful, I mean honestly, that *boy*, he is just a*dor*able! I mean, not that I would rob the *cra*dle, but I seen him in his new little *suit* uptown a while ago, and he reminds me so *much* of Victor Mature! Well, 'bye now, Harry, see you in the funny papers . . ."

Harry, stupefied by the lilac vapors of cologne that swirled about his head, watched her step out the exit door into a perfect fanfare of wolf whistles. Monk's between-the-teeth mating call was distinctive in the shrill cacophony. For a moment Harry felt as if his heart would burst with disappointment and relief.

When it didn't, he edged over to Nurse Ratliff's table and sneaked an unauthorized peek at the *Encyclopedia,* which rewarded him with a memorable object lesson against attempting to judge a book by its cover. It did so by calling itself, on the cover, *The Illustrated Scientific Encyclopedia of Human Anatomy and Sexuality*, by Philander C. Rexroat, Sex.D., and, on the title page, *Prayers for My Good Health*, by P. Cosmo Rexroat, Doctor of Natural Theosophy, Chiropractic Science, and Colonic Irrigation. The text seemed to consist entirely of little meditations with headings like "A Prayer for My Enlarged

Liver," "A Prayer for My Kidney Stones," "A Prayer for My Constipated Condition," and so on—"O Lord, please help Thy servant to comprehend that the Way, and the Only Way, to ease his or her constipated condition is the Naturo-theosophic Way, the Colonic Irrigation Way"—the table of contents a dismal litany of miseries and megrims, a catalogue of irrefutable evidences of God's inhumanity to man. There were, as Rosemary had correctly reported, no full-color plates, nor no Hollywood stars, nor nothing.

"He useta be a peacher," said Nurse Ratliff. She'd come out of the johnny and found Harry looking at the book, but she was past caring. "He was borned with the Gift of Gab. If you get the Gift of Gab, you gotta use it." She sounded resigned, as if the Gift of Gab were a disorder, something he had caught. ("O Lord, please help Thy servant to comprehend that the Way, and the Only Way, to ease his or her Gift of Gab condition is the Naturo-theosophic Way, the Colonic Irrigation Way.")

"Was he ever really *in* the movies?"

"Well, he was in a barbershop quartet, see, and back in the thirties they sung in a Buster Crabbe Western one time. He sung bass, he's the one that went boom-boom-boom."

"How'd he happen to go into sexology?" Harry still hadn't given up the notion that he himself might have a call for the profession—though he did hope to work at a higher level that Dr. Rexroat seemed to have attained.

"Why," Nurse Ratliff said, "he run a tent show in a carnival for a while. Had him a morphadike. That's how he got his start."

Harry decided he'd stick with sportswriting. Mr. Ockerman came downstairs, looking glum, and squeezed himself into the box office, then backed out a moment later with a puny little bouquet of dollar bills in his hand. "Here," he said. "Seventy percent of eleven dollars." He counted out seven dollars and some change onto Nurse Ratliff's table, and pocketed the rest.

Nurse Ratliff tucked away the seven dollars in some dismal grot somewhere inside her sweater. "Listen, hon," she said, "me and him"—she nodded toward the johnny—"is gonna hit the road. We can't sell no litterchoor to these illiterent sons of bitches. Let 'em in and give us the rest of our money, and we'll go on. It's a long way to Cincinnati in that goddamn little cold-ass Crosley. A person's butt sets right on the goddamn blacktop."

Mr. O. received these instructions and observations gloomily, but without debate, merely inquiring how he was to go about returning the doctor's film to them. He was, Harry saw, ready to part company with the august scientist and his winsome colleague on almost any terms.

"You just hang on to it, hon. Him and me are getting out of this line of work anyhoo. You can't make nothing in sex hygiene nowadays, people has got to where they think they're too good for it or something. We're gonna set on our ass and draw his old-age penchant. That print's about wore out, and it's the last one we got. You just keep it for a souvenir."

"I don't want the nasty thing," Mr. Ockerman said. "It's the nastiest old thing I ever seen."

"Yeah," she said amiably, "it is nasty, ain't it?" From the street came the crash of an empty bottle shattering on the sidewalk, followed by a burst of raucous laughter. "But if I was you, I believe I'd show it one more time. Them rubes might turn ugly on you if you cancel out on 'em. Just pitch it out when you're done, if you don't want it."

"What'd I want with it?" Mr. Ockerman muttered. "The nasty thing's fallin' all to pieces anyways." But with two or three fists now beating a heavy tattoo on the outer doors, he saw the logic in the advice. "All right, Harry," he sighed, turning from her, "let the yay-hoos in, then. And you stay close to your light switch while I show it, hear? That old film ain't nothing but splices, it bursted three times during the other show. Shoot, I wish little Nadine hadn't've seen such stuff as that. I hate that worse than anything, a sensitive girl like her."

"I don't think it upset her too much," Harry said.

"Hell," Nurse Ratliff offered, "I seen a lot worse than that by the time I was her age. And it wasn't in no goddamn movie, neither."

"Open up, Harry," Mr. Ockerman said, as he fitted himself into the box office again. "Maybe we can shut our eyes when the nasty parts come on."

SIXTEEN

Harry went on out and threw open the doors. At first there was a little surge of the crowd into the outer lobby, but he saw as he took his post at the ticket catcher's box that the house wasn't going to amount to much—eighteen or twenty at the most. Monk was conspicuously not among them now.

The first ticket went to Elmo Jimson, who'd somehow escaped his mother's iron grip and doubled back on her. He came in with Pinkeye Botts and two or three other boys. Pinkeye showed Harry a half-pint of Sweet Lucy, unopened, inside his coat; he was also carrying a bottle of Nehi orange, for a chaser. "We're gonna get intoxicated as greased owl shit," he vowed as Harry took his ticket. Harry said he believed it.

Kidney Rottington was hanging back in the crowd —because, Harry knew, it would be easier for Harry to slip him in without a ticket if there were no pressing numbers at his back. Rod was on his own these days, the tobacco market having lately summoned The Weave to duty. As Harry and Rod went through their standard make-it-look-good pantomime—in which Rod handed Harry an imaginary ticket, and Harry tore it in two and handed him back an imaginary stub—Harry asked what had become of Monk.

"He took off after that big heifer," Rod said. "We'd bought us a pint together, and when she came out he took off so fast"—Rod giggled happily, and patted his hip pocket—"he forgot all about it."

Harry consoled himself with the thought that it would be a mighty chilly night, even for dicky-dunkin', on Oodles's front porch swing.

As soon as Rod was out of the way, Nurse Ratliff lumbered to her feet, unbuttoned her white smock, and laboriously stripped down to her fur coat and pedal pushers. She wadded up the smock and stuffed it into a shoe box, along with her root-beer apothecary display.

"Hey, Hollywood!" she brayed at the rest-room door. "Did you fall in? Or what?"

Almost instantly the doctor wobbled forth, buttoning up, and struck an attitude of cringing servility before her. "Right-o, thweet-heart," he said humbly. His mouth, sans choppers, had collapsed again like an empty coin purse. It would be a while before he petted another polar bear.

Nurse Ratliff handed him the shoe box, piled her cashbox, a stack of *Encyclopedias*, and the two framed sheepskins in his compliant arms, then took him by his Palm Beach shoulder pads and aimed him toward the exit door.

"Hop to it, old man," she said, impelling him through the doorway with a gentle push. "Let's us make like a sewer, and get the shit outta here."

While the doctor was loading the car, a couple of stragglers drifted in, and at last Mr. Ockerman gave up and closed the ticket window. He came out of the booth a moment later and presented Nurse Ratliff with another little greenback nosegay for her collection.

"You know what?" he said as he handed it to her. "I don't believe you all ever *was* the Cecil B. De Mille of Sex Hygiene Entertainment."

"If we ain't," Nurse Ratliff shrugged, tucking it away, "then who the hell is, I'd like to know." She had a cigarette dangling from the corner of her mouth, and her Red Cross tiara, which she'd forgotten to remove, was cocked over one eye like a pugnacious swabbie's cap.

"I got to patch up that daggone old film again before we start, Harry," Mr. O. said. "It'll be a couple of minutes yet." With the merest twitch of a farewell nod to Nurse Ratliff, he turned and clumped off up the stairs. It was the first time in all of Harry's years with him that he'd skipped his old exultant exit line: "Curtain time, Harry! Curtain time!"

"Well, I'll be switched!" Nurse Ratliff sniffed, looking after him with her fists planted on her hips. "Never said goodbye, kiss-my-butt, nor nothing, the prissy-ass son of a bitch! Hon"—she was addressing Harry now—"never forget your goddamn manners, hear?" He sure wouldn't, Harry promised. "You ain't got any questions about sex or anything for me, have you, sugar?" It crossed Harry's mind to ask whether she happened to know what perversion Mr. Brown's lady friend had practiced, but he decided against it, and said No ma'am, he didn't think so. "Well," she said, "I reckon we'll head out. Don't do nothing I wouldn't do." Harry said he sure wouldn't. At the door she turned one more time, and said, "When I was a girl back in Oklahoma, I knew a boy a lot like you. He had the nicest ways. Butter wouldn't melt in that little bastard's mouth. And, hon"—Harry saw to his amazement that her eyes had misted over —"hon, I thought he hung the goddamn moon." Then she was out the door and gone, and a few moments later he heard the Crosley's little engine start. It sounded like Miss Lute's treadle sewing machine.

The audience, meanwhile, had grown restive. A brief scuffle, ostensibly good-natured, but with a nervous edge, broke out between a Latchburger and a Shanktonian, over some obscure distinction that had excited their respective senses of civic pride. They didn't come to blows, but there was a certain amount of residual growling back and forth between the two delegations. Particularly testy was Scudder Wallingford, a Shankton boy born and reared, and, like the rest of his tribe, of a dependably evil disposition when he was drinking. The show, thought Harry as he hurried into the box office, must go on—pronto.

"Are you *ready* up there?" he inquired anxiously of the genius of the iron blossom—for what was to be the last time for a very long while, though of course Harry didn't know that yet. A pause, then Mr. Ockerman breathed a sigh into the tube so heavy that Harry, all the way downstairs, felt the air stir against his ear. Then at last Mr. Ockerman's "Ready here" followed the sigh down the tube, and Harry threw the houselight switch.

The credits, Harry noted from the seat he'd taken all alone at the rear of the auditorium, made no mention of the illustrious Dr. Philander Cosmo Rexroat, neither as writer nor as producer nor as director—a curious oversight which puzzled Harry at the time and led him, in later years, to speculate that the doctor, in his days as the boom-boom-boom man in the barbershop quartet, had won the footage in a studio poker game, or perhaps found it among the sweepings on the cutting-room floor.

At first, Harry couldn't figure out what Mr. O. had been so exercised about. In the opening scene, Mr. Strong, the stalwart young biology teacher at Centerville High, reminded his students that "there are two kinds of love, class: physical love and lasting, spiritual, sentimental love," and then added, "Uh, all right, class dismissed. Next time we'll take up the alimentary canal." When the gang gathered at Ye Olde Sweet Shoppe after school, for sodas and jitterbuggery, the principal subject of conversation was what a swell guy was Mr. Strong, and how really swell it was that they had such a swell, up-to-date guy to talk things over with. By and large, everything appeared to be perfectly swell in Centerville, U.S.A.

Soon, however, the specter of reaction reared its ugly head. A cabal of Centerville bluenoses took control of the PTA and the school board, and their first order of business was to notify Mr. Strong that if he persisted in his ill-advised efforts to instruct their offspring in the Horrors of Venereal Disease and the Miracle of the Human Reproductive System, he could pack up his advanced Bullshevik ideas and hit the bricks. This transgression of academic freedom instantly plunged all of Centerville back into the Dark Ages, of course, with the natural result that the teenagers began procreating madly and dropping like flies of various social diseases. Eventually—and inevitably—the son of the president of the PTA got the daughter of the chairman of the school board out in his roadster and, under that ovoid harvest moon, committed zygote upon her, and the chickens started coming home to roost. The parents of the unlucky pair suddenly began exerting their influence in more advanced directions, and in no time committees from the PTA and the school board convened on Mr. Strong's doorstep, hats in hand. The remainder of the movie was in fact a movie-within-a-movie, a selection of the visual-aid films with which Mr. Strong enlivened his classes—first the ubiquitous pistil-and-stamen love story; then several 1920's-vintage military-training films indelibly demonstrating the hideous consequences of failing to use the pro-kit after every illicit encounter; and finally that frank and daring foreign film, direct from steamy Stockholm, *The Birth of Swedish Triplets.*

The foreground part of *Dads and Mothers* had obviously been made at least fifteen years ago, the camera work and the sound were often virtually inscrutable, and the film broke four times during the first reel and three times in the other two. But the Burdock County playboys raised a terrible hue and cry whenever the screen went dark, and Harry would dash back and hit the houselights so they could amuse themselves by watching one another's antics while Mr. Ockerman made his splice.

Monk arrived during one of these small crises, sauntering in through the lobby with his heeltaps clickety-clacking as Harry hotfooted it the other way, toward the box office. There was barely time for them to nod in passing, but Monk accompanied his nod with a wide grin and wink, and by the lobby lights Harry could see the grizzle of white angora on the fabled midnight-blue drape-shape sportcoat. A hint of lilac lingered on the air. He'd had, Harry calculated after a glance at the lobby clock, almost an hour to accomplish his libidinous ends. That was more time than Harry had allotted himself for the purpose, but he still wasn't convinced. He just didn't see how they could have done it, not those two behemoths on a porch swing in the cold.

By the time Mr. Ockerman got the film rolling again, Monk had found a seat down front with Rod and the remains of their pint. His arrival created something of a stir, especially when the Toothless One also reported in, so emphatically that Harry heard it back where he was sitting, like the toot of a distant foghorn.

On screen, the enterprising couple had done their discreet vanishing act beneath the roadster's dashboard, and the audience was beginning to realize, to its annoyance, that the camera wouldn't be catching them *in flagrante delicto* after all. The more vigorous critics were already expressing their displeasure in the form of hoots and jeers and their own oral interpretations of the Toothless One's remarks.

The film broke again while the rampant stamens

were in hot pursuit of willing pistils, and now the audience was growing fractious. They threw pennies and trash at the screen, they yowled and scuffled and stomped, they defamed one another, the management, Mr. Strong, and the absent Dr. Rexroat in the most unbridled fashion. While Harry was in the box office attending to the houselights, Scudder Wallingford put his feet against the seat in front of him and, for no better reason than that he felt like it, pushed till the whole row tore loose from its moorings, dumping Elmo Jimson and Pinkeye Botts and their troops on the floor. It made an awful splintering sound, as if the whole place were being pulled down upon their heads.

"What's that, Harry? What's going on?" squeaked the little voice in the iron flower.

"They're getting sorta . . . out of hand, Mr. Ockerman, maybe you could . . ."

"You take care of it, Harry, you're the man in charge down there, I got all I can handle with this daggone old film, here we go now . . ."

On the screen the stamens sprang to the attack once more, and Harry threw his switch with a trembling hand. There were a lot of people in the world that Harry was afraid of, and Scudder Wallingford was among the very foremost. This time he took Mr. Ockerman's long-barreled chrome flashlight with him when he went back to his seat. Its heft alone was in a small way reassuring—and in the dark, the man with a light has the authority. The trouble was that Scudder might be too drunk to know authority when he saw it.

Inside, the clamor had subsided for the time being, the idyll of the pistils and stamens having given way to Mr. Strong's next offering, a slow parade of the walking wounded, those impulsive military personnel (enlisted men only; no WACs or WAVEs, certainly no officers) who had neglected their sexual hygiene, with inevitable results. The footage had no sound track, nor was one needed. Naked and ghastly, rotten with unspeakable diseases, wasted, bejeweled by vile eruptions and excrescences, the specters shuffled one by one before the camera, paused for lingering close-ups of their more ornamental symptoms, and then shuffled off again, some of them grinning at the camera's eye in idiot self-congratulation: *Look, Ma! I'm in the movies!* None looked ashamed, or even very regretful. One shambling wraith even turned and waved a jaunty goodbye, as if to say *I go now to die in the name of love. Is there one among you who would do as much?*

In the back row, Harry sat mesmerized; he couldn't bear to look, yet he couldn't turn away. The apparitions, brain children of the mad scientist Dr. Rexroat, passed before him like the shades of his own darkest fears. So this was what it all came down to! *Omnia vincit amor!*

Harry's reflections were rudely interrupted by

Scudder Wallingford, who was on his feet down front, howling, "Aaah sheeeit, I seen this inna goddamn *army!*"

I'll be killed, Harry notified himself as he advanced reluctantly down the aisle. Near the front, struck by an idea he hoped would prove to be a lifesaver, he thrust his hand into his pocket and found a dollar bill. He took a deep breath and flicked the switch on his flashlight.

"Uh, Scudder," he called in a half whisper, leaning across several patrons into Scudder's row, "if you'd like to have your money back . . ."

Instantly, three or four nearby voices demanded refunds. Harry closed his hand around the dollar and edged his way farther into the row, inadvertently treading on several Shanktonian toes. The grumbling rose alarmingly in volume and vigor. At last, the nervous beam of Harry's flashlight sought out Scudder's face, a mask of indignation in the darkness. "I said I done *seen* this sheeit," he snarled, "inna Goddamn *army!*" Harry turned the light to the dollar in his hand, to show Scudder that the refund offer was still good, and saw Scudder's own hand approach as if to accept it.

Suddenly, not the dollar, but the flashlight was in Scudder's grasp, and in another instant he'd lobbed it end over end toward the screen, the light playing crazily off the walls and ceiling . . . and on the Largest Silver Screen in South-Central Northeastern Kentucky, a colossal syphilitic wretch seemed to look down at himself in horror as the spinning flashlight ripped a gaping black hole in his naked groin and clattered to the floor backstage. Now rough hands laid violent hold of Harry, pushing and hauling him this way and that among an unholy confusion of Shanktonians and Latchburgers. Scudder Wallingford, it seemed, had taken him up as a sort of cudgel, and was belaboring Latchburgers with him . . .

"Hold 'im, Harry!" called a familiar voice. "Here I come!"

"I got him!" Harry croaked absurdly.

Then Scudder flung him away, and he was stumbling over feet, legs, knees, pitching forward, tumbling ass over teacups into somebody's lap, somebody's loose elbow slammed into his temple, somebody's fist flailing blindly in the darkness cuffed his glasses from his face, he banged his chin against somebody's knee and bit his tongue, his mouth filled with the tinny taste of blood, he landed on all fours on the floor amid a shuffling pitch-black forest of legs and feet, his left hand found, caught up his glasses—miraculously intact—even as his right hand was stomped like a spider by a heavy brogan, he cried out in mingled pain and joy, the pale light of the screen flickered overhead, above the riotous clamor he heard Mr. Strong call his class to order and announce the impending birth of Swedish triplets . . .

Spitting blood, his head ringing like a dinner bell, Harry caught hold of a seat back and laboriously picked himself up off the floor. All around him, men had stopped scuffling among themselves to watch a huge rumpus now ensuing in the aisle; two large, lumpy forms grappling, tussling in the gloom, falling back and colliding again . . . It was—Harry managed somehow to locate his face and hang his glasses on it—it was Scudder Wallingford and . . . Mr. Ockerman!

Even more amazing, as best Harry could tell by the faint, fitful light of the screen, Scudder was getting the worst of it! Mr. Ockerman had thrown an admirable hammerlock on him, and was actually half dragging, half pushing him up the aisle, using his own superior mass to excellent advantage. Several times Scudder almost broke away, but each time Mr. Ockerman, working with prodigious strength and nimbleness and tenacity, hung on and drew him back. Without even really thinking about it, Harry understood that Mr. Ockerman had rescued him, and that—because Right sometimes makes Might—his strength was now redoubled.

But Scudder, though squat, was a burly, powerful man suffering a terribly indignity, and he resisted mightily, backing them one step for every two steps they progressed, kicking and flailing with furious drunken abandon, hooking his feet on the aisle seats, digging in his heels, squealing and flopping about as if he were being dragged up from the lower depths not just against his will but against his very nature. Harry shouldered his way into the aisle—the first time in his life he'd shouldered his way anywhere!—and sprang to the side of his employer, where he was instantly rewarded with a Scudder Wallingford elbow to the rib cage that doubled him over and left him gagging for breath.

"I got him now, Harry!" grunted Mr. Ockerman. "I can . . ."

With a fearsome burst of strength, Scudder wrenched free from the hammerlock and, spinning away, went sprawling backwards into the crowd again, bowling over two or three patrons as he fell, and was completely swallowed up by the darkness down between the seats. Harry heard Monk McHorning bellow in pain or outrage somewhere close at hand. Mr. Ockerman also went down heavily in the aisle, but he was unhurt, and immediately began struggling to get his feet under him.

Then Scudder surfaced in the murk brandishing a Nehi orange bottle by the neck, and lunged straight for Mr. Ockerman, who knelt defenseless in the aisle, his bald pate fairly glowing in the dark. With awful clarity Harry saw the glinting bottle arc through the gloom, saw in his mind's eye Mr. O.'s poor beleaguered noggin burst all asunder with the blow, irreparable as Humpty Dumpty . . .

Saw then a hand thrust from the shadows, a huge

white *deus ex machina* hand that caught the bottle in mid-descent and plucked it from Scudder's grasp as blithely as a lady might accept a posy from a lover; saw Monk McHorning step forth and collar Scudder so tightly with his other hand that he almost garroted him with his own undershirt, saw him turn Scudder toward the exit—"All right, shit-for-brains," he said matter-of-factly, "now you done it"—and march him off up the aisle, rapping out a little tattoo on his skull with the Nehi bottle, to keep him stepping smartly.

Scudder wasn't inclined to dawdle, and Harry and Mr. Ockerman caught up with them in the lobby just in time for Mr. O. to hold the exit door open while Monk put his foot on Scudder's rump and propelled him through it, in the general direction of Shankton.

"That knothead stepped on my new shoes," Monk said, dusting his hands. "Which I was forced to wipe it off on his damn ass."

"Listen, young fella," Mr. Ockerman panted. "I want to thank you! I—"

"Aah, fergit it, dad," Monk said affably. He glanced at Harry, and grinned. "I think I owed you one." Turning to go back into the auditorium, he said, "You too, Step." He laughed aloud, and ambled off down the aisle. Beyond him, on the screen, Swedish babies the size of infant hippopotamuses were arriving at a furious rate.

"What's he mean, Harry?" Mr. O. wondered when he'd caught his breath. "He don't owe me anything that I know of."

"He was just talking," Harry said, gingerly dabbing a fat lip with his handkerchief. Privately he surmised that the lilac nimbus and the hoarfrost of angora on Monk's sportcoat probably bore upon the matter. But what did he owe Harry?

"I think it was him that . . . cut one out on my front porch that time," Mr. Ockerman mused. "Maybe he's apologizing."

Harry didn't think so, but he let it pass.

"Well, Harry, we was too much for that old boy!" Mr. O. was expansive, reluctant to let go of their moment of triumph. "I never could've handled him, Harry," he went on, "if you hadn't helt him for me." Harry modestly concurred; he was feeling pretty good himself. Where there had seemed to be no heroes at all a few minutes ago, suddenly there were three—and he was one of them.

On the screen, a proud Swedish father displayed his trio of bouncing baby boys. He looked suspiciously Italian. "Py yumpin' yiminy!" he exclaimed. "Ay ban hit de yackpot!" *Dads and Mothers* was almost over.

"I tell you something, Harry," Mr. Ockerman said, more gravely now. "I'm thinking maybe we ought to just close up shop for a few weeks. Give this old television a chance to shoot its wad, y'know. When people has to set home and watch it all the time, they'll find out it wasn't worth a poot to start with. We got to fix the seats they tore up, and that hole in the screen, maybe we'll just do some painting too, and hang some new drapes and whatnot, and when this television has blowed over . . ."

Afterwards, when the audience was out of the way, Harry and Mr. Ockerman didn't even bother to survey the damage. They got out the stepladder, and Mr. Ockerman held it for Harry while he climbed up and changed the Now Playing sign from DADS AND MOTHERS to CLOSED FOR REPAIRS. Then they turned out the lights and locked up, and Mr. Ockerman assured Harry that the show would certainly go on. Harry had his doubts, but he'd have bitten his poor mangled tongue clean off before he voiced them. Mr. O. said Well, there's no business like show business, is there, Harry? and Harry said Nossir, there sure isn't, and they went their separate ways.

When he reached the car, Harry found a folded piece of paper under the windshield wiper. It was a note, scrawled in pencil on the back of one of the *Dads and Mothers* handbills he had posted around town weeks ago; the corners were torn off, where the tacks had been. The handwriting was Monk's.

"Dear A-hole," it read, "I was force to borow your backseat (watch out for pecker tracks!!) p.s., did you write me my theme yet on What Democracy Means to Me, its due Mon??"

Harry spent the next half hour parked under a streetlight, picking angora fuzz off the upholstery in the back seat. Then he drove home with the windows wide open, to blow out the heady redolence of lilac.

SEVENTEEN

Miss Lute had been feeling poorly for the past several days, but she insisted that Leona take her to church that Sunday morning just the same. While they were gone, Benny—taking a turn at fatherhood that weekend—gave Harry a haircut in the kitchen. As he worked he hummed "On the Sunny Side of the Street" under his breath. He hummed it very slowly, like a dirge; still, it was bouncier than "On the Banks of the Ohio," which was what he usually hummed.

When he was almost finished with the haircut, he said Well, son, we're going to have to do something about your granny.

Miss Lute's present indisposition, Benny intimated, was probably just the beginning of the . . . well, she was slipping, anybody could see that. (Harry couldn't. He thought she'd seemed sharper than ever lately.) They were going to have to make some changes soon, Benny said, that was all there was to it. So last night they'd had a talk with her, he and Leona, and he knew Harry would be glad to hear that they'd ninety-nine percent persuaded her to give up the place and let them all go back to Dayton.

Harry was amazed to discover that this news—if news it was—didn't cheer him. But neither did he let it bother him much, because he was as sure as he could be that Miss Lute wasn't about to move to Dayton; she was stringing them along for some reason.

Miss Lute took to her bed as soon as they got home from church. When Leona had left the house to take Benny to the Greyhound, she summoned Harry to her bedside.

"Harrison Biddle," she snapped, "did that sneak tell you I said I'd go to Dayton?" She was as frail as a kite, the merest bump beneath the bedclothes, but her voice was strong.

"Why, yes, ma'am," Harry owned, "he did." There was no reason to protect Benny's reputation with Miss Lute; it was past saving.

"Well," she said, more softly now, "I ain't a-going."

"No, ma'am," Harry said, laughing. "I didn't think

you would." Then he apprehended what she meant, and wished he hadn't laughed.

Miss Lute paid it no mind. "What do you aim to do with yourself?" she demanded. "You ain't fixing to barber, I hope."

Harry told her he'd been thinking he might be a sportswriter.

"A writer," she said wonderingly. "Well, you always was puny." She pulled herself up in bed. "I want you to take care of something for me, Harrison Biddle. Hand me my knittin'." Harry fetched the knitting bag from her bureau and brought it to her. She fumbled inside, and withdrew a long white envelope, sealed, with Poodaddy's enormous old Sheaffer fountain pen clipped to it. She took off the pen and passed the envelope to Harry, and ordered him to open it. On the flap she'd penciled, in large, shaky letters, "LAST WILL AND TESTIMONY, KEEP OUT!!!" Inside were two sheets of lined tablet paper, neatly folded. They were blank.

"Take that in the front room," Miss Lute said, "and pitch it in the fire."

When Harry came back to her bedside she was chuckling to herself. "Leona Pomeroy thought I'd give it all to the nigger babies," she said. "Lord, Lord, the little snip. I would've went to Dayton with her years ago, just to hush her up—but I didn't want her raising up my grandboy with a bunch of bloomin' Buckeyes."

Somehow, this conversation kept reminding Harry of the one he'd had last night with Nurse Ratliff.

"A writer," Miss Lute said again. "Your Poodaddy would've liked that, he was ever a thoughty kind of man. Here"—she handed him the fountain pen—"this was his best pen. You take it. Ought to be a world of writing in it yet." She folded his hand around the pen; it felt as big as a billy club. Her own hand was all bone and leather, like a bat's wing. She leaned back into her pillow, and Harry could see her trying to focus her rheumy old eyes on him. "Harrison Biddle Eastep," she murmured, "you ain't no old Buckeye, are you, hon?"

"No, ma'am," Harry said. Then he added the words he knew she hoped to hear: "Not any more."

"I didn't think you was," she sighed drowsily, closing her eyes. "Run along, then."

That evening Miss Lute slipped into a coma, and in the night she died.

For the next few days, the world was on its tiptoes. Neighbors came and went on tiptoe, bearing food and sympathy. Preachers and undertakers and lawyers tiptoed in and out like burglars. Benny tiptoed in from Dayton on Monday morning—he'd caught the next bus straight back to Needmore—and his heels didn't touch the floor for days. That evening, after everyone had left, he and Leona tiptoed all over the house in a kind of stifled frenzy, looking in utterly improbable places for Miss Lute's last will and testimony. Harry didn't let on that he had the foggiest notion what they were looking for, until at last Leona noticed the pen in his shirt pocket. Oh, that, he said; Miss Lute gave that to him yesterday, when he burned some papers for her. Leona cried a little then, and called Miss Lute an old sweetheart. Mama *had* changed her mind, she told Benny happily; she burned the will because she'd decided to go to Dayton! It was a reconciliation of sorts, and Harry saw that it would spoil things to tell them what Miss Lute had said, or what was in the envelope. He'd keep that between himself and her. Buckeyes probably couldn't take a joke anyway.

Miss Lute was laid out for visitation at the funeral home on Tuesday evening, from seven to nine. By seven o'clock the place was mobbed; by seven-thirty it was almost empty. Everybody had come, and gone, as early as was consistent with minimal civility. It was the night of the Burdock County–Limestone game.

Only two visitors stayed on after eight o'clock. One of them was Mr. Ockerman, who sat with Harry for a long time on a sofa across the room from Miss Lute's casket, telling him how much they would all be missing him around there, but how if a person was intending to broaden himself and expand his horizons and whatnot, he couldn't find a better place to do it in than Dayton, Ohio. Harry appreciated the encouragement, though he knew that he was a son of Needmore now, and that would never change.

Still, you could never tell: Buckeye girls might go for an adventurer like him, home from the hot climes.

The second lingering mourner was the farmer who'd bought Poodaddy's land from Miss Lute five years ago, after Benny left. Now he wanted the house, too. He and Leona and Benny sat till almost nine-thirty in the furthermost corner of a parlor just off the visitation room, their heads together, talking in hushed voices. By the time the man got up to go, they'd struck a deal. Harry, waiting alone with Miss Lute in the other room, could almost hear the gnashing of her choppers.

He lay sleepless for a long time that night, buffeted by alternating waves of anticipation and remorse. Finally he dropped off, hoping Miss Lute would soon be so pleased with her own change of scenery that she wouldn't begrudge him his.

EIGHTEEN

Miss Lute's funeral was on Wednesday morning. Harry was a pallbearer, along with Benny and one of Miss Lute's distant cousins and four of Poo-daddy's old courthouse cronies; once again he found himself in the company of men.

Afterwards, the house was full of people for a couple of hours, paying their respects. Harry stayed back in the kitchen as much as possible, trying to teach Pittybiddle to say "Luticia Pomeroy Biddle." But either the old bird didn't like Harry or he was grieving, or both; he wouldn't peep.

Somewhere during the course of the morning, Harry heard what had happened in the Limestone game. For the first three quarters Monk had been a terror; by the half he had nineteen points, and in the third quarter he racked up nine more; the Bulldogs led, 36 to 24. The crowd was delirious: they hadn't beaten Limestone since 1941. Then, on the first play in the final quarter, Monk was called for charging. Still holding the ball, he professed his innocence. The referee reached for the ball, and Monk said something to him. The referee turned to the scorer's table and hollered, "Technical!" Monk said something else, and the referee hollered, "'Nother technical!" Monk said something else, and the referee hollered, "*'Nother* technical!" And that was when Monk turned around and chucked the basketball straight through an open window and out into the night.

He was instantly ejected from the game, of course, and was reported to have left the floor with a huge grin on his face. Norval Stroud replaced him, and the Bulldogs never scored again. Limestone won 40 to 36, going away. Monk was in disgrace, a dark stain on the winning tradition of the Bulldogs.

That afternoon Harry took the Hudson over to school to pick up his books and arrange for the office to transfer his records to Dayton. He and Benny were leaving tomorrow, Benny to go back to work, Harry to start school. Leona would come later, when the estate was settled.

He cleaned out his locker and loaded his books into the car. Then he went to the office. Normajean Stickler took down the address of the Dayton school and said, as best he could make out through the Viennas and saltines, that she'd send his records out as soon as she could get around to it. As he was leaving, she called him back and said "the Doctor" wanted to see him in the inner office.

Nobby was at his desk, gnawing on a pencil and spitting out the pieces, in the throes of a black and wrathful sulk.

"Eastep," he declared vehemently, "you can lead a horse to water, but there ain't a way in the world you can make him drink."

Harry acknowledged that that was his understanding too, and always had been.

"The boy has let us down, Eastep. He has let down ever single one that offered him the helping hand of friendship. He has let you down, that helped him with his studies so he could better hisself. He has let Claude Craycraft down, that kindly went and bought him a suit of clothes out of his own pocket so he'd be dressed nice, like the other boys. He has let the fans down, and he has let the school down. He has let me down, myself personally. He has let the Bulldogs down, Eastep. He has bit the hand that feeds him." Nobby held his own hand up before him, demonstrating invisible toothmarks; he looked as if he'd like to bite Monk back. "Eastep," he concluded bitterly, "it's my opinion the boy is uncorrigible."

It was a conclusion Harry couldn't argue with. "Well, coach," he chirped, his heartiness so transparent that even Nobby could see through it, "I want ya t'know it's been a real pleasure workin' with ya, and . . ."

"They tell me," Nobby broke in—and as he spoke he spun his swivel chair to the blank wall behind his desk, so that Harry suddenly faced the back of his bald head—"that you'll be leaving us, Eastep."

It took Harry, standing there with his mouth open, a full quarter of a minute to realize that they had just bid each other goodbye, and that he could go now.

"Hurry back now, Eastep," said Normajean with a spume of cracker crumbs as he went out. Harry didn't answer; he'd always hated being called his last name by a lady.

In the hall, he ran into Ramona Halfhill and a couple of other Doggettes on their way to cheerleader practice. Ramona stopped long enough to say she'd heard that he was leaving, and was real sorry to see him go. "You was getting to be a whole lot better kisser, Harry," she said in parting, "and I mean that sincerely."

He found Monk in the second-floor boys' room, as he had known he would; it was sixth period, time for Current Events, and Monk always spent Current Events in the second-floor boys' room. He had thrown the window wide open, and was sitting on the sill, cleaning his fingernails with a kitchen match.

"Hey, here's Step'!" he said cheerfully. "Take yer hat and jacket off, Step! Did yez get yer granny planted?"

"Yeah," Harry said, "we did. I hear you got in trouble."

Monk laughed. "Gimme a butt, Step. I gotta start buyin' my own now, Craycraft cut me off. He is *pissed*, dad!"

Harry shook two Kools from his pack. "What'd you say to that referee, anyway?" he asked after they'd lit up.

"When he comes up to me to get the ball, I sez Hey, chuck you, Farley, I never charged that guy. I had decked the poor son of a bitch, see, but I sez Hey, I never charged that guy. So he calls a technical on me. You tell 'em, Horseshit, I sez, you been over the road. He calls another technical. You tell 'em, Sycamore, I sez, you got the fuzzy balls. And he calls another damn technical. Then I seen that window, and that was all she wrote, dad!" Now Monk paused and sobered and spoke more confidentially, though they were still alone. "I done it on purpose, Step. I was looking to get throwed out."

Agape, Harry said he didn't get it.

On the way home from Lexington the other day,

Monk said, Claude Craycraft had got to talking basketball. Only he wasn't talking wins and losses, he was talking point spread. He was talking I scratch your back and you scratch mine. He was talking suits of clothes and money in your pocket. Monk had understood right away what he had in mind, and—because he liked a new sportcoat now and then as well as the next man—he had known immediately that he was going to do it. They'd keep on winning, Claude explained, as often—well, almost as often—as they could, only not by quite as many points, sometimes, as people maybe thought they might. But of course what they needed first were a few big wins, to build up the confidence of the sporting public. Most of all, what they needed right then was a big win over Limestone.

But wait a minute, Harry said. Then why did you . . .

"Listen," Monk broke in, "I like to put the ball in the hoop, see, I mean I like to play the damn game. And last night I was rollin', Step, I had my moves and I was rollin'! Twenty-eight damn points in three quarters! And then I think to myself Hey, one of these days, dad, you're gonna be rollin' along nice this way, you got your moves and you feel fine, see, and all of a sudden you'll have to start screwin' up, kickin' it around. You'll have to turn into just another Burdock Bulldog, dad! Because you work for that damn old lizard Craycraft. Or else keep on puttin' it in the hoop—and work for the Nobster! And that's when I throwed it out the window. I figured I just as well let both of 'em know they'd better not be depending on me.

"So," Monk said after a pause, "you're leavin'."

Harry said he was. Everybody's haulin' ass, Monk said. Did you hear about the Sharpies? Harry hadn't.

"That damn Kiddington run off with The Weave's old lady."

Amazing. The lovely June, Madonna of the People's Bank of Needmore, had traded in one Sharpie for another.

"Some people," Monk said, "hasn't got principle one. You can't count on people these days, Step. It don't pay." He took a meditative drag on his Kool. "I'm cuttin' outta here myself in a couple of weeks," he said.

He was going to join the army, he told Harry. After a few more games like last night, the Nobster would be glad to sign his papers for him, testifying that he was seventeen—though he wasn't even sixteen yet—and that he had parental consent.

"How do you know he will?" Harry couldn't help asking. "He's pretty pissed himself, maybe he won't do it."

"Because if he don't," Monk said, "he's got a big hungry orphan boy on his hands. Which this orphan boy," he added, laughing, "does not play no roundball whatsoever."

Harry said he got the picture: Don't mess with the Well-Built.

"Damn straight," Monk said. "Craycraft wanted his damn sportcoat back. People in hell wants ice cream too, I told him."

Harry could see the Hudson in the parking lot below their window. From the cafeteria, where the cheerleaders were practicing, wafted the frail, desultory chant: "*California oranges, Texas cactus! We play Crutcher County just for practice!*" Sixth period was drawing to a close. He told Monk he guessed he'd better be on his way.

"Well, I hate to see you go, dad. I was kinda hopin' I'd get me another chance to use your back seat before I cut out. But you shake it easy now. Don't let yer meat loaf. And hey"—he settled himself more comfortably on the windowsill—"it's been real, dad!"

Moments later, as Harry crossed the parking lot, Monk whistled to him from the window. "Hey, Step," he called, "if you see Kay, tell her I love her."

EPILOGUE

On an unseasonably warm afternoon in mid-June, the Southbound Long Dog labors along Kentucky 10, a twisting two-lane blacktop that snakes through the hilly northern Kentucky farm country. The destination of the bus, according to the sign above the windshield, is Lexington.

There are, in addition to the driver, two passengers aboard: one is a garrulous old woman in the seat directly and immediately behind the driver; although the driver has not so much as grunted in response, she has been talking to him about her grandchildren since the minute the bus pulled out of the station in Cincinnati, where she got on. She is on her way to Tennessee, to visit the vaunted grandchildren.

Across the aisle, in a window seat farther back in the compartment, sits the second passenger: Harrison Biddle Eastep, M.A.

This is not the Harrison Biddle Eastep he once was. This model is upwards of twenty-five years old; he has filled out some, and his skin has cleared up. His hair is newly cut, too high in the sideburns; not Benny's work. He wears a shiny, pond-slime green wash-and-wear suit of a positively reactionary cut—narrow shoulders, three buttons, pipestem arms and legs—so new it still smells faintly of the chemicals of which it was fabricated. No drape shape for Harrison Biddle Eastep, M.A.

The M.A. is almost as new as the suit, the haircut, and the simulated-leather attaché case in the seat beside him. In the attaché case are a clean shirt, a few toilet articles, and a bound copy of his master's thesis: *Cosmic Optimism and the Theory of Historical Periodicity in Nineteenth-Century British and American Thought,* Submitted in Partial Fulfillment of the Requirements for the Degree of Master of the Arts, Department of History, Ohio State University.

A few weeks ago, this Harry Eastep was a shambling, shaggy-headed graduate student on the GI Bill, a proto-crypto-pseudo-semi-quasi-Beatnik in blue jeans and cycle boots and his old army fatigue jacket. In the proto-crypto-pseudo-semi-quasi-Beatnik salons of Columbus, where he and his graduate-student friends sat on the floor on grass mats and drank cheap wine and listened to *The Sounds of Sebring* at top volume on the hi-fi, he sometimes held sway with one of his famous Freudian interpretations of this or that episode in the course of human history—theories which his friends found fascinating (though perhaps less so than he supposed), and which his professors found so repellent that, overcompensating, they often awarded them higher grades than their limited originality would otherwise have merited. Propounding these illuminations as a grizzled veteran of the United States Army (a reluctant draftee, he'd put in most of his two years as a clerk-typist at Fort Leonard Wood, Missouri), Harry was listened to and even modestly admired among his peers, especially by Buckeye girls of a more adventurous sort—the intense ones, who went in for Camus, black leotards, and no makeup. Still . . . they weren't the Draw-Me girl.

But all that's behind him now. This Harry Eastep, the newly minted Master of Arts on the bus, is a job seeker. He is presently en route to Lexington, where he is to have an interview on the following morning with a Dr. Furlong, chairman of the History

Department of Transylvania University, to discuss his qualifications for an appointment to the position of instructor in Dr. Furlong's department.

This enterprise, unfortunately, is doomed to fail; for it will develop that Dr. Furlong—a deeply repressed individual, in Harry's opinion—is cool to the point of outright hostility toward Freudian readings of history. Later that summer, after a number of similarly unsuccessful interviews, Harry will be hired, by mail, to a position at a state teachers' college in Oregon.

Of course, Harry doesn't know that disappointment and Oregon are in his future, and for now his heart is filled with hope. Since graduation—only ten days ago, but already it seems longer—he has lived with Leona in Dayton, in the modest tract house she and Benny bought years ago with Miss Lute's estate. Benny ministers to the sideburns of the angels now; one afternoon, when things were slow at the Orville Wright, he climbed into his own chair, dozed off, and never woke up. Leona makes a very decent living trading in small commercial properties, a calling for which she has, just as she'd long suspected, something of a gift. Laundromats, filling stations, neighborhood grocery stores, even a poolroom now and then—those are her stock-in-trade. She continues to read her Samuel Shellabarger and Frank Yerby, but otherwise has little use for history, which she regards as insufficiently modernistic; she cannot imagine, she declares, why any son of hers would want to piddle around with all that old-timey stuff. Nonetheless, for graduation she gave Harry the slime-green suit, the attaché case, the haircut . . . and two months to find employment.

Sighing, he takes his thesis from the attaché case and settles back in his seat for a bit of boning up, in preparation for Dr. Furlong. As he opens the thesis, he allows himself a moment to enjoy its heft, the satisfying crackle of its spine, the double-barreled title. He does experience a twinge of regret that his cranky old curmudgeon of a committee chairman wouldn't permit him to explore his theory of the American Revolution as an Oedipal manifestation, but all in all, the taking up of the thesis cheers him enormously, as it always does.

In a few more miles, he reminds himself, the bus will pass through Needmore. He doesn't want to be so engrossed in his reading that he forgets to look out the window at it.

> It is ever the way of the antiquarian [Harry reads] to pick through the flotsam and jetsam of the past until he brings to light the tiniest, least significant detail or fact or coincidence in the whole mountainous heap of the fossilized detritus sloughed off by great events and Higher Truths, and then solemnly to proclaim himself the discoverer of a microcosm. Thus do minor

functionaries come to be venerated as the powers behind great thrones, minor skirmishes as the turning points of great wars, minor poets as the seminal influences upon great epics; thus do small-minded men, studying human experience through the wrong end of time's telescope, persuade themselves that they too have a certain significance in the grand scheme of things . . .

He sleeps. Lulled by the drone of the engine and the warmth of the afternoon and the rocking motion of the bus and the voice of the tedious grandmother and the turgidity of his own prose, he falls sound asleep before he's read a page. He awakes as the bus pulls to a stop in front of an exact replica, in about two-thirds scale, of the Burdock County courthouse.

"Well sir," the grandmother is chortling, "if that youngest one ain't the *beatin'est* little scamp . . ."

The bus driver stands and stretches. "Needmore," he rasps. Harry, groggily agog, struggles to inflate the courthouse to its original dimensions. Now the driver is taking a roll of newspapers—*Enquirers*—from the overhead rack, just ahead of Harry's seat.

"Excuse me!" Harry hears himself call. "Excuse me, if I get off here, can you give me a refund?"

The driver, a beefy, florid man with a disfiguring squint, nods, inaffably. "Ain't another bus through here till tomorrow, buddy," he warns, giving the back of his hand to the empty, sun-washed streets around them. Harry immediately catches his drift: Buddy, he is saying, there ain't nothing here that need detain the traveler.

"That's okay," Harry says. "I'll hitch the rest of the way." He's on his feet now, stuffing the thesis back into his attaché case. He cannot imagine why he's doing this.

The bus driver shrugs, and counts out the refund into Harry's hand. "I used to live around here," Harry apologizes, following the driver off the bus. "I thought I'd look around a little."

"I'll tell that youngest one I seen you," the jolly granny calls after him. "He's a buster, that'un is!"

The driver crosses the street and drops the roll of papers on the doorstep of—yes!—Conklin's drugstore. Like the courthouse, indeed, like everything Harry can take in at first glance, Conklin's is precisely as he recalls it, but diminished, a scaled-down reproduction of itself.

"Where is everybody?" Harry wonders aloud when the bus driver comes back across the street. It's an ordinary weekday afternoon, but Needmore looks like Sunday; everything seems to be closed, and there are only two or three parked cars the whole length of Courthouse Street.

"It's Wednesday," the driver says. He swings himself up through the folding door, then turns and squints down at Harry, as though the green suit offends his eyes. "A lot of these little shit-ass towns close down on Wednesday afternoon, nowadays."

The door slaps shut, and seconds later the bus pulls out, through a cloud of its own noxious vapors. And there, in the middle of Courthouse Street, with the last thin, blue shreds of exhaust fumes still aswirl about him, stands Harrison Biddle Eastep, M.A., thinking Well, ain't *this* a hell of a note!

For a moment he feels dizzy, and faintly nauseated. It passes, his land legs come back to him, and Needmore slowly begins to assume its proper proportions. He sees that a few things have changed, after all: what used to be OK Package Liquors is a blackened, burnt-out ruin, and what used to be the butcher shop has become OK Package Liquors; Blankenship's has become Blankenship & Sons; Bertha's Beauty Box has become Mr. Shirley's Beauty Box. Craycraft's appears to have been closed for years; its only window is covered with a weathered sheet of plywood, and the neon "Billiards" sign dangles over the sidewalk by a single strand of wiring.

Just off the far end of the block, the New Artistic stands its lonely sentinel. Harry steps back into the street for a better angle on the Now Playing sign:

CLOS FO EPAIR

S

Now, he tells himself as he approaches the theater, maybe he'll find out why he got off the bus.

The New Artistic's doors are as snugly padlocked as he and Mr. Ockerman had always left them, but some disgruntled patron of the arts has heaved a piece of concrete block through the plate-glass pane of the entrance door, so that Harry has merely to step over the sash—gingerly, so as not to snag his suit or scratch his attaché case on the jagged shards of glass—and he is in the outer lobby.

The floor is littered with broken glass, bits of fallen plaster, whiskey bottles. Someone has wrenched the iron grill of the ticket window partway from its frame, and the wooden panel below the window has been kicked in, smashed to flinders. On the wall where they used to post the Coming Attractions, the tattered corner of an old poster announces "Roy," the name formed by a coiling lariat. Underfoot, another torn poster declares "Movies Are Better Than Ever."

And the light is wrong, there's too much of it. The door to the inner lobby stands wide open, and daylight, broad daylight, is streaming through. Harry advances, steps over a heap of debris—he recognizes his old ticket catcher's box, a cornucopia of ticket stubs—into the inner lobby. From the head of the aisle he sees how the sunshine has invaded this place that never saw the light of day for forty years: the roof is gone.

Not gone; collapsed. That constellation of clumsy stars has given way at last and crashed to earth, taking with it a good part of the stage, the first twenty rows of seats, the very floor itself. The rear third of the auditorium, protected by the overhanging balcony, is intact; the seats are still there, patiently awaiting the audience that never comes. But where the balcony ends, so does the aisle. Harry edges down till the floor creaks ominously beneath him. The falling roof has sheared the floor off cleanly all the way across the room; at Harry's feet a chasm yawns, and at the bottom of the chasm is a huge snarl of mangled tin roofing, broken timbers, ruined seats, hanks of savaged velveteen. The wall beyond it, where the stage should be, is festooned with long, pale tatters of the Largest Silver Screen in South-Central Northeastern Kentucky. Above it all is a great square of open sky, and afternoon sunlight spilling in.

Now a Carolina wren, no bigger than a mote in Harry's eye, appears atop the west wall, dances briefly along the narrow parapet, then glides down the shaft of sunlight, catches, clings to a shred of the tattered screen, and bursts into ringing, joyous song: *Tea kettle, tea kettle, tea kettle, tea!*

Harry thinks of Pittybiddle, who, following Miss Lute's example, pined away and died in Needmore before Leona could get him moved to Dayton. If the wren were Pittybiddle reincarnate, what name would he call? Luticia Pomeroy Biddle? Harrison Biddle

Eastep, M.A.? The thought is not altogether disquieting; there is the suggestion of continuity in it, and of some distant peace. The Son of Needmore, standing here in the midst of this blasted grandeur in his Philadelphia lawyer getup, has come home. He hasn't been here in the longest time, and for all he knows he'll never be back again; but he's here now, and this is home.

"Mister," says a small but authoritative voice close behind him, "you better get your tail end outta here!"

He turns. Standing at the head of the aisle is a five-foot-tall, one-hundred-pound miniature of Monk McHorning.

"Who says so?" Harry asks, somewhat less forcefully than he'd intended. Unlike the other edifices in Needmore, this Monk McHorning replica refuses to become life-size; but it's impressive even in miniature.

"None of your damn beeswax," the boy replies. He is nine or ten, roughly the age Monk would have been when the redheaded lady with the fox furs visited the orphans' home. His black eyes glitter like obsidian chips, and his prognathous little jaw juts as combatively as the original's. Harry has to remind himself that, as a full-grown Master of the Arts, it behooves him not to be intimidated by a ten-year-old boy.

"My mama says so, by God," the boy goes on. "This

here is her show hall, and she don't want nobody a-messing in it."

"Hers?" Harry asks. "What happened to—?"

"She heired it from my grandpa. He went and died."

"Listen," Harry says, "it's okay. I used to work here. Is Nadine Ockerman your mama? I used to work here, for your grandpa." He decides to take a flyer. "Hey," he adds heartily, "I used to know your daddy!"

"My mama's name ain't Ockerman," snaps the boy. "It's Worthington, by God. Her name is Nadine Worthington, and my name is Newton Worthington, and my daddy's name is—"

"Shirley Worthington," Harry finishes. He should have known; Mr. Shirley.

"He'll pin your damn ears back for you, bud," warns Newton Worthington, without much conviction. "He'll jerk a knot in your tail." His scowl is as black and menacing as he can make it; he does not look remotely like a Newton Worthington.

"Well," Harry says, moving up the aisle, "don't worry, I won't be but a minute longer."

The boy reluctantly backs off before him. "I woultn't let him catch me in here if I was you," he says from the outer lobby. "My daddy's a bad cat to clean after."

"Don't I know it," Harry assures him. "I'll just be a minute."

At the outside door, the boy pauses and says, addressing not just Harry but the looming ruins about them, "I don't want nothing to happen to this show hall, see." He steps over the sash of the broken door to the sidewalk, where he has parked his bicycle. "I'm gonna put me on some shows in here one of these days," he vows, mounting the bike. "I seen a picture in a magazine of a show they was putting on in the Radio Musical City Hall. When I get big, I'm gonna fix this old show hall up, and put me on one like that, by God." He wheels the bike about and tears off up the street.

Alone again, Harry sets down his attaché case and goes back into the inner lobby. The door to the box office hangs from one hinge; he wrestles it aside and enters the little booth. At first he thinks the speaking tube is gone. Then he sees that it has been torn loose from the wall and twisted upward, like the tendril of a living plant turning its bloom to the light. He pulls it down and brings it to his lips.

"Are you ready up there?" he whispers urgently. "Are you *ready* up there?"

Though he waits a long time, no answer is forthcoming. But that doesn't really matter, Harry tells himself—for the show goes on, because it must.